Contents

1　Evolutionary Processes

The most cursory examination of the living world is sufficient to reveal that organisms are wonderfully adapted to the environments in which they live. By the early nineteenth century it was gradually becoming realised that the harmony of organisms and their environments need not necessarily be explained in terms of special creation. However, it was not until the publication in 1859 of Charles Darwin's *On the Origin of Species by Means of Natural Selection* that the theory of evolution was systematised and laid upon firm foundations of evidence. Darwin's theory has been modified and refined during the last century, but its essentials stand unchallenged by serious thinkers.

Natural selection

The basic mechanism of evolution is *differential reproduction*. Such familiar phrases as 'the survival of the fittest' and 'the struggle for existence' do not quite fit the situation, because what is of evolutionary importance is not survival *per se*, but is rather reproductive success. Those organisms which are evolutionarily *fit* are those which propagate themselves most successfully.

Natural selection is the name applied to all those factors which tend to eliminate or promote individual genes or gene combinations in the gene pool of a population. Evolutionary change is made possible because all individuals within a species are genotypically and phenotypically unique, so that within a population there is a continuous range of variation in all characteristics, from adaptively inferior to adaptively superior. Most selection is in fact *centripetal*, or selection towards the population mean, i.e. working to eliminate mutations which occur; the selection which is of evolutionary importance, however, is that which determines that gene frequencies in the filial generation will, however slightly, differ from those of the parent generation. Evolution thus consists of the gradual modification through time of the gene pool, and hence the array of phenotypes, of a population. Such modification takes place at any one time for one of three classes of reason: either the animal becomes better adapted to a stable environment (this is a relatively minor and restricted type of evolution); or it changes in response to environmental change; or it adapts to a new environment into which it has moved. It is usually found that evolution is comparatively rapid subsequent to the adoption of a new habitat, those animals best adapted to the new conditions being very distinctly favoured selectively. Features initially evolved in one ecological setting, but which turn out to be particularly advantageous in a new way of life, are known as *preadaptations* to the latter, although this of course in no way implies any sort of predestination. A rule of thumb, though by no means a natural law, is that evolutionary changes are irreversible. Finally, it should be re-emphasised that the unit of evolution is not the individual, but the population. Selection may eliminate individuals, but it does not change them.

One way in which gene frequencies may vary from one generation to the next in the absence of natural selection is known as *genetic drift*. Briefly, in a very small population, and only in such a population, gene frequencies may vary over time purely by the agency of chance. But although genetic drift is usefully invoked in explaining certain inter-population variations, its effect on long-term evolutionary change is negligible.

Parallelism and convergence

Similar selective pressures may cause quite unrelated animals to evolve strikingly similar features. Hence both bats and insects possess wings, although obviously this could hardly be used as a criterion of relationship. Two terms frequently encountered in this context are *parallelism* and *convergence*. Dr G. G. Simpson has defined parallelism as: 'the development of similar characters separately in two or more lineages of common ancestry and on the basis of, or channelled by, characteristics of that ancestry', and convergence as: 'the development of similar characters separately in two or more lineages without a common ancestry pertinent to the similarity but involving adaptation to similar ecological status'. In general, parallelisms are said to occur between animals sufficiently closely related to be placed in the same order, while convergence is said to take place between orders or even higher categories.

Two further terms of great importance are *homology* and *analogy*. Homology refers to characters of different animals which resemble each other because they are derived from a common ancestor, while analogues are characters which resemble each other purely because they represent responses to similar functional demands. Thus the wings of bats are merely analogous to those of insects, while the hands of monkeys and men are homologous because they are

Man's Ancestors
AN INTRODUCTION TO PRIMATE
AND HUMAN EVOLUTION

Also in this series

The Microstructure of Cells Stephen W. Hurry
The Protozoa Keith Vickerman and Francis E. G. Cox

Ian Tattersall holds M.A. and M.Phil degrees from Cambridge and Yale Universities respectively and is at present doing research at the Peabody Museum of Natural History at Yale. He is the author of several papers in various areas of physical anthropology.

Cover illustrations
Front Female long-haired spider monkey, *Ateles belzebuth*.
Photo by author
Back Female sifaka, *Propithecus verreauxi verreauxi*, with infant, photographed on the Mangoky River, western Madagascar.
Photo by author

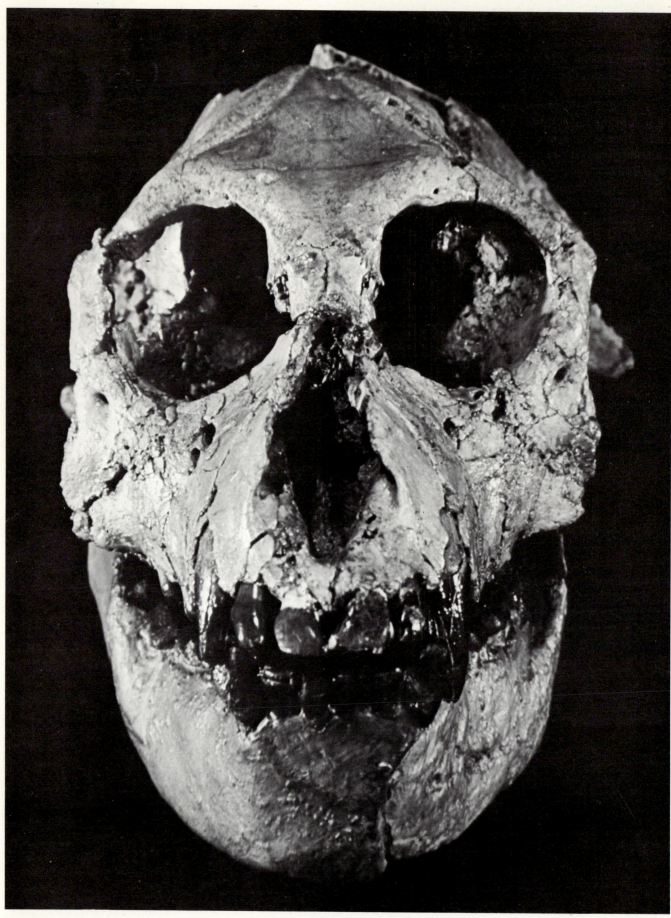

Facial view of *Aegyptopithecus zeuxis*, a primitive ape from the Oligocene of North Africa, some 27–28 million years ago. The cranium and mandible are not associated, and the anterior part of the mandible is reconstructed.

Courtesy of Dr E. L. Simons

Introductory Studies in Biology   General Editor: Stephen W. Hurry

Ian Tattersall

Man's Ancestors

AN INTRODUCTION TO PRIMATE AND HUMAN EVOLUTION

John Murray   Albemarle Street   London

Preface

Palaeoanthropology is undergoing a period of rapid change. The introduction of new techniques of investigation and the recent large augmentations of the hominid fossil record have both contributed to this. Yet in spite of the fact that we are now able to recognise many earlier misconceptions as such, we are in one sense no closer to understanding the course of human evolution than we were a decade ago. Far from clarifying the relatively simple concepts then held, the new fossil finds have demonstrated that the hominid evolutionary picture is of far greater complexity, and its details further from our grasp, than was previously imagined. On the other hand, new approaches to the interpretation of the fossils have enabled us to arrive at a plausible and consistent model of hominid differentiation, even if this—dietary specialisation—is less flattering to the modern ego than were most former hypotheses. In this book I have attempted to indicate the new directions of investigation and to provide as clear a picture as possible of the course of man's evolution as it is presently understood, although I have not tried for the sake of simplification to present any clear-cut schemes where none exist.

For reasons of space I have had to omit discussions of some topics which ideally should have their place in a book such as this, and to restrict discussion of others. For instance, the reader will find almost no account of the history of the study of man's ancestry and little detail of archaeology, while a knowledge of elementary genetic terminology has been assumed. I hope, however, that most readers will find my choice of material reasonably satisfactory.

For whatever may be useful in this short text I owe far more than the conventional debt of gratitude to my teachers and friends David Pilbeam and Elwyn Simons. Thanks are also due to Peter C. Ettel, M. F. Gibbons and my wife Elisabeth who read the manuscript; to Al Coleman and Diane Barker for their photographic assistance; to Rosanne Rowen, Susan Weeks and Carl Wester for the artwork; and to all those, separately acknowledged in the figure legends, who kindly allowed me to use illustrations of their own.

New Haven, 1969   I.T.

Printed in Great Britain by Jarrold & Sons Ltd, Norwich

0 7195 2188 2

characters of common ancestry, a fact attested to by their fundamental identity. They do of course differ in details of size, shape and proportion, but they are built on the same basic plan, representing variations on a common ancestral theme. The ability to distinguish between homology and analogy plays a crucial part in the palaeontologist's business of reconstructing the evolutionary history and relationships of the animals he studies.

Speciation

Among sexually reproducing animals, the *species* may be defined as the largest group of animals capable of inter-breeding and producing fully viable and fertile offspring. The population units within which interbreeding is *likely* to take place, however, are almost invariably much smaller than the full species group. Such population units are known as *demes*. A number of demes in turn comprise a *subspecies*, a group occupying a distinct geographical area, and possessing adaptations to local environmental conditions. Species containing more than one subspecies are known as *polytypic*.

The words 'fully-fertile offspring' comprise a vital part of the species definition. Among the primates, there are many examples of interspecific mating. For instance, the baboon species *Papio anubis* and *P. hamadryas* live in adjacent territories, and hybridise in the zone of contact. The species remain distinct, however, because the hybrids are less fit than their parents; the probability of a hybrid's reproductive success is less than that of his half-brother whose parents are both of the same species. Again, it is important to remember that the species definition applies to the behaviour of animals in the wild. Many adaptations are behavioural, and the conditions of captivity invariably distort natural behaviour. Under laboratory conditions, many monkey species can be made to interbreed and produce offspring fully fertile with either parental species. In the wild, however, these species do not interbreed with each other; behavioural isolating mechanisms or geographical separation prevent it. The rare reported occurrences of natural interspecific matings are invariably due to abnormal circumstances.

Speciation is the process whereby a single parent species gives rise to one or more daughter species. Among mammals, speciation is invariably *allopatric*; that is, it takes place when a population becomes isolated by a geographic barrier from the other populations comprising the species. In isolation, the genetic compositions of the two groups thus formed will gradually diverge as they adapt to local conditions without the exchange of genetic material. If the barrier persists sufficiently long, accumulated genetic changes will be sufficient to render hybrids less fit than members of the parental populations. When, on withdrawal of the barrier, interbreeding is again possible, hybridisation may take place initially, but if hybrids are of reduced fitness, isolating mechanisms will rapidly evolve to eliminate the waste of reproductive potential which hybridisation involves.

The multiplication of species in this way is one of the two fundamental processes of evolution, and is known as *cladogenesis*. The other primary evolutionary process may be referred to as *anagenesis*, the gradual evolutionary change a lineage undergoes through time. As evolution occurs in a lineage, descendants will gradually become less and less like their ancestors, until the point is reached where an ancestral population at time A is sufficiently dissimilar from its descendant population at a later time B to warrant specific distinction.

The scientist who deals with fossils faces a far more complex situation than does his colleague who studies living animals. First, he obviously cannot use interbreeding criteria in deciding whether a collection of fossils represents one, or more than one, species. Secondly, the evidence with which he deals is usually sparse, fragmentary and widely scattered in both space and time. Animals look alike because they belong to the same species, but the palaeontologist must start from the other end, allocating fossils to the same species if their differences are within the range of variation which one would expect to find in a genetic species. Species groupings based purely on morphological similarity are known as *morphospecies*.

Palaeontological morphospecies, because they exist in the dimension of time as well as in that of space, can, in the context of time, be regarded as *time-successive species*. A lineage of animals, consistently isolated reproductively from other species, will gradually change through time until the morphological differences between populations occurring earlier and later in the lineage are as great as, or greater than, those observed between living species. The lineage contains no species boundaries within itself; such boundaries as we wish to draw are merely arbitrary, for convenience of reference, and correspond to no natural unit. It has been aptly pointed out that any attempt to divide an evolutionary continuum ultimately involves drawing a boundary between a father and his son! However, fortunately what is most important is not the division of a lineage, but rather its delimitation and the definition and interpretation of the evolutionary trends discernible within it. The almost invariable incompleteness of the fossil record frequently provides places at which division may conveniently be made, in the form of time-gaps for which no fossil material is available, but this merely defers the problem until such time as the gap in our knowledge is filled. With the realisation that such divisions are arbitrary and made purely for the sake of convenience, it is obvious that the best criterion for division is one of time, since it is, at least theoretically, perfectly objective, although only too frequently the dating of fossils is equivocal.

Mosaic evolution

Different parts of an animal may evolve at different rates. Thus the early forerunners of man had man-like teeth but small brains, and the foot appears to have approached its modern form somewhat before the pelvis did. Of course, any animal is an extremely complex integrated unit, so that its components cannot change entirely independently of each other, but this phenomenon, known as *mosaic evolution*, is repeatedly demonstrable in the fossil records of all animal groups, including man's.

Adaptive radiation

In two basic evolutionary situations, an animal stock may undergo ecological diversification within a relatively short time period. The first of these follows the development of a new adaptive type; the second is when an animal group adopts a new territory, either geographical or purely ecological. Such diversification is known as *adaptive radiation*, and may be brief or prolonged in time, and ecologically wide or narrow. The impetus for this diversification is of course provided by the simultaneous presentation to all animals involved of a new spectrum of adaptive zones. It is this simultaneity, together with the fact that the radiation is derived from a single original adaptive type, that distinguishes adaptive radiation from the progressive occupation of new ecological zones, the other major component of diversification.

2   Classification

The classification of living organisms, an essential step towards their understanding, is based on the system devised during the eighteenth century by the great Swedish naturalist Carolus Linnaeus, and published in his book *Systema Naturae*. The definitive edition of this work, the tenth, was published in 1758. Linnaeus's scheme is hierarchical; that is, it consists of grouping species into successively more inclusive categories. Although Linnaeus, predating Darwin, did not believe in evolution, the hierarchical system is an apposite one in the classification of living animals, whose relationships are most conveniently expressed in terms of their phylogenetic closeness (recency of common ancestry).

Before we discuss the mechanics of classification, it will be as well to define a few terms which any student of evolution must invariably encounter. The broadest of these terms is *systematics*. This refers to the study of the kinds and diversity of organisms, and of the relationships among them. *Classification* is the process of ordering animals into sets on the basis of these relationships, while *taxonomy* is the name given to the study of the theory, procedures and rules of classification. *Nomenclature* applies to the naming of each of the groups comprising a classification, and its rules are clearly set down in a publication entitled the *International Code of Zoological Nomenclature*. An animal is named according to an objective procedure, but it is classified according to the subjective judgement of a scientist, and its place in a classification is always subject to the possibility of change.

Classification starts with the species, the only natural category, and the unit which evolves. Subsequent more inclusive categories are arbitrary, but we hope that the groups we establish do reflect 'real' units in terms of *phylogeny*, or evolutionary relationship. From the level of the Order (the largest unit in which we are interested here) down, the more important of these categories are listed below:

Order
Suborder
Infraorder
Superfamily
Family
Subfamily
Genus
Species

Species are grouped into genera, similar genera into subfamilies, similar subfamilies into families, and so forth. Of course, the degree of 'similarity' involved is at each successive stage more remote. The names applied to superfamilies, families and subfamilies are given special endings: *-oidea*, *-idae* and *-inae* respectively. Thus the chimpanzee, a member of the genus *Pan*, belongs in the superfamily Hominoidea, the family Pongidae and the subfamily Ponginae. A useful term in describing these categories is *taxon* (plural: *taxa*), which may be used of a unit at any level of the hierarchical scale.

What has been said above implies that classifications ought to reflect phylogeny; perhaps this statement should be modified slightly. One pole of modern taxonomic thought is represented by the *pheneticists*, who hold that classifications should be based purely on morphological similarity. The most vociferous of this group are the numerical taxonomists, who argue that since most characters are polygenic, and most genes pleiotropic, the quantitative consideration of large numbers of characters will provide a sampling of the genotype sufficient to enable one to produce a classification based on genetic similarity. This approach has provoked a great deal of very valid criticism from the *cladists*, or phylogenetic taxonomists, who represent the opposite pole of taxonomic opinion. Phylogenetic classification in its strictest form depends on the time of separation of ancestral forms. Thus the degree of phylogenetic relationship may be expressed as follows: a species x is more closely related to species y than to species z if, and only if, it has an ancestral species in common with y that is not at the same time the ancestor of z. A classification based on such premises can, however, be misleading. Thus, it is not beyond the bounds of probability that man and the chimpanzee share a common ancestor which was not the ancestor of the gorilla. But it would be ludicrous to classify man and the chimpanzee together in a group separate from that of the gorilla. Obviously, the genotypic and phenotypic differences separating man and the chimpanzee are far greater than those between the chimpanzee and gorilla, and any classification ignoring this would be confusing. The fact that evolutionary change may take place in different lineages at different rates is the primary reason why pure cladistic classification is not always completely satisfactory.

Ideally, the method of classification employed in any given case should depend on the purpose for which the classification is made. In nearly all instances, however, the most satisfactory classification is one based on phylogeny, but modified where necessary to avoid cases such as the one just discussed.

Every species is identified by a *binomen*, which is a combination of two italicised Latin names. Thus, modern man comprises the species *Homo sapiens*. The first component of the binomen is the *genus*, or generic, name, while the second is known as the *trivial* name. The two together compose the species name. Every species must have a *type* specimen, or *holotype*, the individual to which the name is attached, and on the identification of which the species name must stand or fall. This does not imply, however, that the holotype of a species is 'typical' of that species. Since species are grouped into genera, there may be several species bearing the same generic name; thus in the genus of the gibbons, *Hylobates*, there are several species named, for instance, *Hylobates lar*, *H. concolor*, etc. Similarly, species in different genera may share the same trivial name, e.g. *Australopithecus africanus* and *Dryopithecus africanus*. The combination, however, is unique. If a group of fossils is assigned to a new species, and is subsequently discovered to be merely a variant of a species already named, the species name wrongly applied to the new material is dropped and becomes known as a *junior synonym* of the earlier name, which henceforth applies to the whole group.

Finally, perhaps it is appropriate at this point to clarify two terms which are frequently misused: these are *primitive* and *specialised* (or *advanced*). These can only apply with reference to a specified lineage or taxon. In any lineage, those characters which are primitive are those appearing early in its evolutionary history. Specialised characters are those evolved later. In any taxon, the primitive characters are those more or less resembling those of its ancestors, while characters which have evolved away from the ancestral type are advanced.

3 The Time-scale of Primate Evolution

Since any fossil represents a point in an evolving continuum, it is essential that we be able to assign it to its correct place in that continuum. In other words, we must be able to date our finds.

Relative dating

The oldest method of dating fossils depends on the assignment of the deposits in which they were originally buried to their correct position in the sequence of geological events. Rocks are grouped into three major categories, but the only kind with which we need concern ourselves here are sedimentary rocks, those composed of compacted and cemented particles weathered from pre-existing rocks and transported to their places of deposition by water, ice or wind. Such rocks comprise about three-quarters of those exposed at the Earth's surface, and are the only ones which contain fossils. The science dealing with the sequence of sedimentary rocks is known as *stratigraphy*. One of the major problems encountered by stratigraphers is the correlation of the sequences of rock layers in different places, since rock type rarely stays uniform over large areas. Temporal correlation is usually accomplished where possible by the recognition of marker fossils known to be characteristic of a given period of time. The incomplete nature of the stratigraphic and fossil records, and the fact that the widespread migration of a form evolved in a single place takes an indeterminate amount of time can, however, make surmise as to the time-equivalence of strata in different areas a hazardous business.

Stratigraphers use a number of different terminologies to identify the different kinds of units with which they deal. *Rock units* are strata of distinct composition, and are the exposed portions of local sedimentary sequences. These are the real, objective units with which the stratigraphers work in the field. More abstract, *time-stratigraphic units* include all rocks deposited during a given period of time. The basic time-stratigraphic unit is the geological *system*. Systems are subdivided into *series*, which are in turn composed of two or more *stages*. These time-stratigraphic units correspond to the completely abstract *geological time units*, which are those to which we will most frequently refer. Systems correspond to geological *periods*, series to *epochs*, and stages to *ages*. Thus, for instance, the rocks of the Tertiary system are those deposited during the Tertiary period.

The sequence of geological time units during the last 70 million years (the time during which primates are known in the fossil record) are shown in Fig. 3.1.

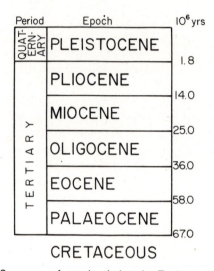

Period	Epoch	10^6 yrs
QUATERNARY	PLEISTOCENE	1.8
TERTIARY	PLIOCENE	14.0
TERTIARY	MIOCENE	25.0
TERTIARY	OLIGOCENE	36.0
TERTIARY	EOCENE	58.0
TERTIARY	PALAEOCENE	67.0

CRETACEOUS

FIG. 3.1. Sequence of epochs during the Tertiary and Quaternary periods, with approximate dates. Primates may have differentiated towards the end of the Cretaceous period, the final period of the Mesozoic era.

Absolute dating

Many naturally occurring atoms (those referred to as radioactive) possess unstable nuclei which spontaneously decay to a state of lower energy. When a 'parent' atom decays, it changes to another type of atom known as the 'daughter' product. The rate of decay of such nuclei is effectively independent of external conditions, and is characteristic of the particular kind of atom. It is not possible to state the length of time it will take for all parent atoms in a system to become completely converted to daughter atoms—this is in theory infinite—but it is relatively easy to express the rate of decay of radioactive parent atoms in terms of their *half-life*, or the time required for half the parent atoms in a system to decay. Geochemists have used these properties of radioactive atoms to provide a means of dating rocks containing them.

There are two basic approaches to such *radiometric* dating: the *accumulation* method, based on the accumulation of daughter atoms formed by the decay of a radioactive parent, and the *decay* method, based on the loss of parent atoms within a system. A variety of techniques, using both approaches, has been developed for dating various kinds of rocks and minerals, of which the two elaborated below are at present of the greatest significance for students of human evolution.

The potassium-argon (K-Ar) method

The radioactive potassium isotope potassium-40 is present as 0·0122% by weight of all natural potassium. Potassium-40 decays in two ways: 89% of the atoms become calcium-40 by a process known as beta decay, while the remaining 11% goes to argon-40 by electron capture. In theory, this provides two complementary methods of dating, but since it is not practically possible to determine the amount of calcium-40 daughter product originally present, the potassium-40–calcium-40 method is not much use in practice. The K-Ar method is an accumulation technique, and shares with related procedures two theoretical assumptions: that the system has remained closed, i.e. that the system has not exchanged any parent or daughter atoms with its environment; and that there were no daughter-type atoms trapped in the system when it formed. The amount of argon-40 in a sample is determined by melting the sample under vacuum and analysing the amount of argon gas which is given off. The age of the sample is then derived by comparing this amount to the known isotopic abundance of potassium-40 in natural potassium, and applying the *decay constant* derived from the half-life.

Argon, an inert noble gas, never becomes chemically bound into newly formed potassium, but may be trapped mechanically in some minerals. Hornblende, biotite and muscovite, however, rarely incorporate argon during crystallisation, and measurement of the argon-40 present in these nearly always provides reliable dates. Volcanic rocks are especially suitable for dating by K-Ar because they are precise stratigraphic indicators and because they tend to have simple thermal histories: that is, they crystallise at high temperatures at which no mineral can retain any argon, so that when the rock cools it will contain no argon-40. Any argon-40 that we then measure is certain to have been formed by radioactive decay. When rocks are deeply buried, the pressure exerted by material above them generates heat, so that argon-40 may be driven off subsequent to initial cooling. In such cases, the date derived will represent a minimum date, since it represents the last time the rock cooled below about 300 °C.

Since the half-life of potassium-40 is relatively long $(1·3 \times 10^9$ years), the method is suitable for dating very old rocks, although it has been used to date rocks as young as 100 000 years.

The carbon-14 (C^{14}) method

The ages of young deposits are most frequently measured by the carbon-14 method, a prime example of the decay clock. C^{14} (radiocarbon) is produced from nitrogen-14 in a neutron reaction, governed by cosmic ray influx, in the upper atmosphere, and decays with a half-life of about 5730 years. This half-life is not that used in practice to calculate C^{14} dates, however, since for many years the half-life was believed to be 5570 years, and calculations were made on that basis. To avoid confusion, dates are still calculated on the short half-life, with the recognition that all dates will be low by about 3%. New radiocarbon becomes incorporated into the world carbon cycle, with which all living things are in equilibrium. When an organism dies, it becomes isolated from the cycle, and the C^{14} it contains at that time starts to decay, while the content of other carbon remains constant. The age of a carbon-containing sample can then be measured by deriving the ratio of C^{14} to other carbon present. Since the half-life of C^{14} is relatively short, when a sample is above about 50 000 years old, the ratio of C^{14} to other carbon is so low that it cannot accurately be measured, and although various isotope enrichment procedures have been applied in attempts to extend the range of C^{14} dating to around 70 000 years, 50 000 years can generally be regarded as the effective maximum of its range.

Errors

Absolute chronometric dates are usually accompanied by ranges of error when they are published. Thus a C^{14} date might read: 5000 ± 300 years, or a K-Ar date: 25 million $\pm$ 2 million years. This is because the same analysis performed on the same material a number of times may yield a series of slightly differing results. A number of conventions may be followed in assessing error, but it has been admitted that in the present state of the science, intuition plays a major role in this. Nevertheless, on a date derived under well-controlled conditions, error will be small relative to the length of time measured, though it must be remembered that the limits of confidence on a date increase sharply as a chronometric method approaches the maxima and minima of its range of usefulness.

4 The Order Primates

Primates is perhaps the most varied and diverse Order of mammals, which makes it one of unique fascination but at the same time one of the most difficult to define. Its lack of diagnostic characteristics led Sir Wilfrid Le Gros Clark to characterise primates instead on the basis of a number of evolutionary trends, among which are: retention of a generalised limb structure and dentition; increasing digital mobility; replacement of claws by flat nails; development of stereoscopic vision and reduction in the organs of smell, leading to shortening of the face; increasing time of dependence of offspring on mother; and progressive development of the brain, especially in the cerebral cortex. Not all these trends are discernible in every primate line, but the list does serve to give a good indication of the general adaptive similarities which make the Order an unquestionable natural unit.

One of the most striking features of living primates is that they display a series of forms exhibiting successively higher levels of organisation which in a sense approximate various stages in primate evolution. This *scala naturae* provides us with a great advantage in our attempts to reconstruct the anatomy and behaviour of extinct primates, including the precursors of man. Such a concept must not, however, be taken to imply that primate evolution progressed inexorably in a straight line towards man, simply throwing off groups at various stages along the line as it progressed. The idea that evolution proceeds along some predetermined course towards an inevitable goal is called *orthogenesis*, and is completely fallacious; evolution is opportunistic, not programmed.

The conventional classification of living primates is given in Table 4.1.

All living primates fall into a cohesive natural group, but at the subordinal level there are two distinct groups, Prosimii and Anthropoidea. The two suborders share the same fundamental features, but exist on different levels of organisation. The *prosimians*, which have survived not greatly changed from Eocene primate stocks, are generally small, with relatively smaller brains than those among the Anthropoidea, or *higher primates*. Their behaviour is more stereotyped, more innate, than is that of higher primates; this is particularly evident in the fact that their capacity to react to unfamiliar situations is limited, while most members of Anthropoidea show high degrees of inventiveness.

Any detailed discussion of the living primates is impossible

TABLE 4.1.
CLASSIFICATION OF LIVING PRIMATES

ORDER PRIMATES
  Suborder Prosimii
    Infraorder Lemuriformes
      Superfamily Lemuroidea
        Family Lemuridae
          Subfamily Lemurinae
          Subfamily Cheirogaleinae
        Family Indriidae
          Subfamily Indriinae
        Family Daubentoniidae
    Infraorder Lorisiformes
      Superfamily Lorisoidea
        Family Lorisidae
          Subfamily Lorisinae
          Subfamily Galaginae
    Infraorder Tarsiiformes
      Superfamily Tarsioidea
        Family Tarsiidae
          Subfamily Tarsiinae
  Suborder Anthropoidea
    Infraorder Platyrrhini
      Superfamily Ceboidea
        Family Cebidae
          Subfamily Callimiconinae
          Subfamily Aotinae
          Subfamily Pitheciinae
          Subfamily Alouattinae
          Subfamily Cebinae
          Subfamily Atelinae
        Family Callithricidae
          Subfamily Callithricinae
    Infraorder Catarrhini
      Superfamily Cercopithecoidea
        Family Cercopithecidae
          Subfamily Cercopithecinae
          Subfamily Colobinae
      Superfamily Hominoidea
        Family Hylobatidae
          Subfamily Hylobatinae
        Family Pongidae
          Subfamily Ponginae
        Family Hominidae

12

in a book of this length whose primary concern is human evolution, but in order to place man firmly in the context of the animal world, a brief discussion of the forms most closely related to him is necessary.

The lemuriformes comprise the most fascinating and diverse infraorder of prosimians. Isolated on the island of Madagascar since the Eocene, the lemuriformes have undergone a remarkable adaptive radiation, evolving a wide variety of arboreal and terrestrial forms. Tragically, many of the most interesting and advanced Malagasy primates have become extinct since the relatively recent arrival of man on the island. The surviving lemuriformes are grouped into three families of which the largest, Lemuridae, contains two subfamilies, Lemurinae and Cheirogaleinae. Lemurinae contains three genera: *Lemur*, *Hapalemur* and *Lepilemur*.

FIG. 4.2.   *Lemur catta*.

FIG. 4.1.   *Lemur fulvus sanfordi*.

The first two of these are diurnal and gregarious, while *Lepilemur* is nocturnal and tends to a solitary existence. There is an interesting variety of locomotor behaviour in this subfamily; *Lemur* is primarily quadrupedal, i.e. all four limbs play approximately equal roles in locomotion, while *Lepilemur* is a vertical clinger and leaper. Vertical clingers and leapers prefer vertical supports for resting; they leap between such supports by pushing off with the hind limbs, on which they also land. Such animals are characterised by hind limbs which are much longer than their forelimbs, and which are capable of extremes of flexion and extension; their feet are long and possess powerful grasping big toes. *Hapalemur* possesses a locomotor pattern which includes many characteristics of both vertical clingers and quadrupeds, as, indeed, does one species of *Lemur*, *L. catta*. Thanks to the studies of Dr Alison Jolly, the social behaviour of *L. catta* is well known, and shows that 'higher primate' intelligence is not a prerequisite of typical primate social organisation, in which learned behaviour plays a crucial role.

The subfamily Cheirogaleinae consists of three small, nocturnal genera: *Cheirogaleus*, *Phaner* and *Microcebus*. The largest cheirogaleine is *Phaner*, an extremely agile quadruped which leaps a great deal. The social behaviour of *Phaner* is little known, but it probably lives, like *Cheiro-*

FIG. 4.3.   *Microcebus murinus*, the mouse lemur: the smallest Malagasy primate.

Courtesy of Dr John Buettner-Janusch

galeus, solitarily or in pairs. *Microcebus*, possibly the smallest primate, darts around in low trees and bushes in a rather squirrel-like fashion, and is usually found in pairs, though like many other nocturnal primates, it may form larger groups for sleeping.

Indriidae contains three genera: from largest to smallest, *Indri*, *Propithecus* and *Avahi*. All three are vertical clingers and leapers, the first two diurnal, *Avahi* nocturnal. The social behaviour of *Propithecus* has been studied by Dr Jolly, and turns out to be in essence reminiscent of that of *L. catta*, although the group size of *Propithecus* is smaller, averaging around five individuals, as compared to a dozen or more.

The other lemuriform family is Daubentoniidae, containing the single species *Daubentonia madagascariensis*, the aye-aye. This is the strangest Malagasy prosimian, and is very little known. Its anterior teeth are very much like those of rodents, and its posterior teeth are reduced to pegs. *Daubentonia* is solitary, and its diet is primarily insectivorous.

The infraorder Lorisiformes contains one family, Lorisidae, which consists of two subfamilies, Lorisinae and Galaginae. The Asian lorisines are represented by *Loris* of Ceylon and southern India, and *Nycticebus*, which occupies parts of South-east Asia. Both are nocturnal inhabitants of dense tropical rain-forests, and both are slow climbers. Slow climbing is a very deliberate quadrupedal mode of progression, the foot being placed behind the hand of the same side before the hand is released to take the next step. Leaping never occurs. This locomotor pattern is most likely an adaptation to bird-stalking. Its practitioners have long, slender limbs of approximately equal length, and powerful grasping extremities with the thumb and big toe set at 180° to the axis of hand and foot. *Loris* and *Nycticebus* are thought to be primarily solitary, though they may occasionally form pairs.

FIG. 4.4.   Propithecus verreauxi.

FIG. 4.5.
Daubentonia madagascariensis,
the aye-aye.
        From D. G. Elliott, 1913

FIG. 4.6. *Loris tardigradus*, the slender loris of Ceylon.
© *Zoological Society of London*

FIG. 4.7. *Galago*, the bush-baby.
Courtesy of Charles D. Miller III

The African lorisines, *Perodicticus* and *Arctocebus*, are also slow climbers, solitary or pair-forming, arboreal and nocturnal. The single genus *Galago*, the bush-baby, of the African subfamily Galaginae, is a vertical clinger and leaper, and on the ground shows this in a bipedal hopping mode of progression. *Galago* is more social than the other lorisiformes, and forms sleeping groups of six up to nine animals.

The final prosimian infraorder, Tarsiiformes, contains the single genus *Tarsius*, the tarsier, with three species, all occupying South-east Asia. *Tarsius* has enormous eyes, a concomitant of its nocturnal life, and has been considered by some to have attained a level of organisation higher than that of other prosimians. However, most of its 'advanced' characters appear to be correlated with its great

visual development, and it seems pointless to remove *Tarsius* from Prosimii. The name *Tarsius* derives from its greatly elongated ankle region, a specialisation associated with its vertical clinging and leaping locomotion. Little is known of the behaviour of the tarsier in the wild, but in captivity up to one-third of its activity is concerned with marking and maintaining territorial boundaries. Tarsiers are usually found in pairs, but males, and females with young, may be found alone.

The higher primates fall into two basic divisions, separated at the infraordinal level: Platyrrhini, the New World monkeys, and Catarrhini, Old World higher primates. These two groups appear to have arrived independently at the higher primate grade of organisation. The New World

FIG. 4.8. *Tarsius*, the tarsier. Although shown here sitting on a horizontal branch, these animals commonly prefer vertical supports.
© *Zoological Society of London*

motor categories. This is especially so in the case of Atelinae, which contains three genera: *Ateles*, *Brachyteles* and *Lagothrix*. The atelines are basically quadrupedal, but possess many locomotor characteristics usually associated with more specialised groups; they are, in fact, frequently referred to as 'semi-brachiators'. Certainly, they do show many signs of forelimb dominance. *Ateles*, in particular, uses its arms a great deal in swinging among the branches and in raising and lowering itself. It also walks bipedally and spends much of its time sitting with its trunk held erect. The morphology of *Ateles* reflects this emphasis on the arms, its skeleton possessing many of the features associated with full-time arm-swingers. It is difficult, however, to say which of these are concomitant to trunk erectness rather than specifically due to the animal's suspending itself by its arms. Its arms are slightly longer than its legs (the reverse is true of most specialised quadrupeds) and its long, hook-like hands lack thumbs. Its chest is broad but shallow, and its spine is set more deeply into the thoracic cavity, thereby bringing the centre of gravity of the trunk closer to its supporting member; its spine is relatively short, stout and inflexible, as opposed to the long, springy vertebral column of quadrupeds, and its scapulae (shoulder-blades) lie at the

FIG. 4.9.   The golden lion marmoset, *Leontideus*.

FIG. 4.10.   *Ateles*, the spider monkey.

monkeys of South America, characterised by the possession of three premolar teeth in each tooth row, as opposed to the two of catarrhines, are divided into two families. The more primitive forms, the marmosets and tamarins, are grouped into the family Callithricidae. These diurnal animals live in small family groups in the high forest canopy, where they adopt a quadrupedal springing form of locomotion.

The more advanced family, Cebidae, is very diverse, but all its members are diurnal except for the large-eyed *Aotus*, the night monkey. The most common social unit is the small family group, though in forms such as the very vocal *Alouatta*, the howler monkey, troops may be much larger. In two subfamilies, Atelinae and Alouattinae, the tail is prehensile and tactile, and is used as an 'extra hand' in locomotion.

Locomotion among the cebids is very varied, and demonstrates very well the artificiality of creating discrete loco-

back of its thorax rather than at the sides. But because leaping is still an important activity, the hind limbs of *Ateles* are by no means undeveloped. Morphologically and behaviourally, *Ateles* may be interpreted as representing a stage intermediate between conventional quadrupedalism and arm swinging.

The Old World higher primates include the Old World monkeys, the apes, and man. The Old World monkeys, forming the superfamily Cercopithecoidea, are divided at the subfamilial level into Cercopithecinae and Colobinae. Geographically, there are also two groups, African and Asian, with only the genus *Macaca* common to both areas, but the taxonomic and geographical divisions do not coincide exactly.

The colobines are less abundant than the cercopithecines in terms of numbers of individuals, but very little so in terms of numbers of species. They are completely vegetarian, their

FIG. 4.11.   *Colobus guereza,* the white-tailed colobus monkey.
© *Zoological Society of London*

diet being primarily of leaves, to cope with which they have evolved highly complex stomachs, and are mostly arboreal, though some species do a fair amount of ground-living. Colobines are all quadrupedal though some, for instance the proboscis monkey, *Nasalis*, may occasionally arm-swing.

There is considerably more variety in habitat among the cercopithecines, which accounts for the fact that the behaviour of members of this subfamily is better known than that of the colobines. This is because several cercopithecine lines have taken to living in woodlands or open country, where they are easily accessible to observation. Another reason for the widespread interest in the terrestrial forms is that they occupy a niche similar to that occupied by early man. It is thus perhaps inevitable that there should have been a tendency amongst some workers to overemphasise the usefulness of inferences from these primates as to the behaviour of early hominids.

The most intensively studied cercopithecines are the savanna-dwelling baboons of the so-called 'cynocephalus group' of the genus *Papio*. These animals live in effectively closed social units which vary greatly in size, but commonly contain around 60 individuals. Such troops consist of adults of both sexes, plus juveniles and infants: they usually contain more females than males. Troops live in fairly large home ranges from which they never stray; within the home range there is a 'core' area containing favourite feeding and sleeping trees. The home ranges of different troops may overlap, but core areas never do. There are distinct differences between males and females in size, morphology and social role; the females are much smaller than the males, have smaller canine teeth, and are much less aggressive. Primates on the ground are much more vulnerable than in the trees and the baboon troop on the savanna is therefore a tightly knit defensive group. Aggression within this group is minimised by the maintenance of a rigid dominance hierarchy. Normally, dominance is linear, i.e. one animal is dominant to another in all circumstances, but occasionally two lower-ranking individuals may band together in a particular situation to dominate another animal normally dominant to both. Positions in the dominance hierarchy are established by play fighting while the animals are juvenile and unlikely to inflict any serious harm on each other. Fights are normally ritualised and harmless, and when one looks dangerous, it is broken up by dominant adult males.

Female baboons go through an oestrus cycle, and are only sexually receptive at around the time of ovulation. As they come fully into oestrus, females will mate with progressively more dominant males, until at its peak, when ovulation actually occurs, they form 'consort pairs' with dominant males. Such a pair will retire to the periphery of the troop for a few days until the female is impregnated. Females with young form a focus of attention for the group, and mother and infant will stay constantly together until weaning. Thereafter the offspring will spend an increasing amount of time with its peers. Learned behaviour is an extremely important part of a primate's total behavioural repertoire, and the mother-offspring relationship is essential to the process of learning. Laboratory tests have demonstrated that infant monkeys deprived of contact with their mothers show severe psychological maladjustment, and are unable to fulfil conventional social roles.

The behaviour of savanna *Papio* was long considered to be typical. Recent studies by Dr Thelma Rowell on forest-living baboons have suggested, however, that this is not the case, and several other lines of evidence imply that *Papio* is a recent immigrant to the savannas. The forest-living baboons observed by Dr Rowell, living in areas of greater food abundance and lessened danger of predation, have a much looser social organisation than their savanna-dwelling counterparts. There is less tension in the troop, which is not a closed social unit, and no dominance hierarchy is apparent. There is an important lesson to be learned from this: the effect of environment on behaviour. Since so much of the behaviour of primates is learned, their behavioural repertoire is particularly sensitive to ecology.

Baboons of the species *P. hamadryas*, living in very arid areas of Ethiopia and Somalia, show yet another form of

FIG. 4.12a.   Baboon troop watching python in Amboseli National Park, Kenya.   *Courtesy of Dr T. T. Struhsaker*

FIG. 4.12b.   *Cercopithecus aethiops*, the vervet monkey.
Courtesy of Dr J. S. Gartlan

social organisation. The typical hamadryas social unit is the one-male group, consisting of a single adult male plus a 'harem' of from one to four females, with offspring. The females are very jealously guarded by the male, and never stray very far from him. At night a number of such groups may gather at sleeping-cliffs, but their cohesiveness still persists. This type of social organisation is thought to correlate with the harsh conditions under which the hamadryas live; scarce food resources severely limit the number of individuals which can live in a given area, so the reproductive potential of the species is enhanced by a form of social organisation which minimises the number of males and maximises the number of females.

One of the most interesting cercopithecine genera is *Theropithecus*, the gelada baboon, today surviving only in isolated parts of Ethiopia. Forming social units similar to those of the hamadryas, geladas are the most terrestrial of the baboons, and exist on a diet of small morsels, such as grains and roots, which they put directly into their mouths with their fingers. This has had a most interesting effect on their dentition. Their anterior teeth, which are little used, are small, while their cheek teeth, which are specialised for grinding, have large, flat crowns, and wear rapidly. Because geladas spend most of their time sitting, the sexual skin of the female, which swells during oestrus and acts as a sexual signal, is located not around the genitalia, as in other cercopithecoids, but on the chest. A curious analogy to hominids!

In the superfamily Hominoidea are grouped the lesser apes (gibbons and siamangs), the great apes, and man. Each of these groups is best regarded as deserving familial distinction.

The gibbons and siamangs (family Hylobatidae) are the only primates which indulge in true *brachiation*. This is a mode of locomotion in which the arms alone, extended above the head, are used to suspend the body and propel it through the branches. Gibbons also climb quadrupedally, and may walk bipedally along branches or on the ground, holding their elongated arms behind their heads. Brachiation is a spectacular, but not particularly efficient, mode of locomotion; perhaps it is best viewed as a feeding adaptation, permitting the animal to feed on fruits growing on the outermost branches of trees. Morphological correlates of brachiation among gibbons are similar to those described for *Ateles*.

Uniquely among non-human hominoids, gibbons form permanent pairs, plus offspring, and they viciously defend a small (quarter of a square mile) territory. Males and females are of about the same size, and, unusually, both sexes possess long, projecting canine teeth.

The rarest of the great apes, *Pongo pygmaeus*, the orangutan, is restricted to the tropical rain-forest of Sumatra and Borneo. As this type of habitat has shrunk, so has that of the orang, and persistent hunting of these animals now threatens their extinction. Their depletion has affected their social organisation, but it has been suggested that this might originally have resembled that of forest-dwelling chimpanzees. Orangs are frugivorous and virtually never come to the ground. Their locomotion is quadrumanous, i.e. both hands and feet are used in grasping supports, and most frequently takes the form of slow, deliberate climbing, though they often hang by their long arms from their long, hooked hands.

The social organisation of the chimpanzee, genus *Pan*, is known from a number of studies. These animals inhabit tropical forest, open woodland and forest fringes, and move with equal facility on the ground and in the trees. On the ground they occasionally walk or run bipedally, but they are primarily *knuckle-walkers*. Knuckle-walking is a quadrupedal locomotor pattern in which the forelimbs are supported on the backs of the middle phalanges of the fingers. In the most terrestrial Old World monkeys, the patas monkey and gelada baboon, which are digitigrade quadrupeds, a loss of grasping power has occurred as a result of the terrestrial specialisation of the hands, and these forms are therefore indifferent arboreal climbers. Knuckle-walking, on the other hand, in permitting the retention of long, flexed fingers, allows its practitioners to travel on the ground or in the trees with equal ease. In the trees chimpanzees hang and swing by their arms, clamber around quadrupedally, or even knuckle-walk or move bipedally along branches. Forest-living chimpanzees spend more than three-quarters of their time feeding in the trees, while those occupying open woodland spend most of their time on the ground, and their diet is more varied, being supplemented occasionally by the killing of small antelopes, bush pigs or colobus monkeys. Both groups, however, sleep in the trees, where nests are built anew each night.

For a long time it has been thought that chimpanzees form open groups of fluid composition, but recent work by the Japanese suggests that the troops seen by observers may in fact be subgroups of a much larger social unit whose composition does not change. The most basic of these subgroups is that of females plus offspring, although troops of widely varying composition are found. Dominance seems to play a very minor role in most social interactions; and any male may mate with an oestrus female.

In many ways we may discern among chimpanzees elements of behaviour reminiscent of that of humans. Off-

FIG. 4.13. Hylobates, the gibbon: captive animal at Yerkes Regional Primate Research Center, Atlanta, Georgia.
Courtesy of Rosanne Rowen

FIG. 4.14.   Two juvenile chimpanzees grooming. Note the knuckle-walking posture of the right hand of the animal on the left.

© *Zoological Society of London*

spring stay with their mother for up to five years, far longer than among any other primate apart from man, and during this period they absorb a great deal of learned behaviour. After this period of dependence, chimpanzees still maintain contact with their mothers, and it frequently transpires that adult females who spend a lot of time together are mother and daughter. Even adult males, which tend to wander around far more than do females, occasionally return to visit their mothers. Such reunions are accompanied by elaborate displays of affectionate greeting. This demonstrates the fact that family ties are recognised even when the animals involved spend most of their lives apart. As would be expected in a promiscuous society, the father-offspring relationship is not recognised. Matings probably do not occur between mothers and sons; the incest taboo may not be restricted to humans! Indeed, it appears that when adolescent males begin to mature they tend to leave the maternal group to join another, returning only occasionally thereafter. If this is so, it may plausibly be interpreted as a mechanism to decrease the likelihood of mother-son mating.

Chimpanzees are highly intelligent animals, and laboratory experiments have shown that, like humans, they tend to solve problems by means of hypothesis-testing, rather than by trial and error. A very interesting feature of chimpanzee behaviour, involving foresight, has been observed in Tanzanian woodland chimpanzees by Jane van Lawick Goodall. At certain times of the year, termites grow wings and cluster in tunnels close beneath the surfaces of their mounds. At such times, woodland chimpanzees will pick and trim grass stalks or thin twigs, and set out to find termite nests in this condition. When one is found, the chimpanzee scrapes away the surface of the mound until a tunnel is revealed, and then pokes the stick down it. After a moment, it withdraws the stick, to which termites are clinging, and picks these off with its lips. This behaviour, known as 'termiting', has been viewed as an example of the making and use of tools, since the chimpanzees are taking natural objects,

modifying them to suit their needs, and applying them to a foreseen purpose. A further feature of chimpanzee behaviour which can be interpreted as tool-making exists in the use of chewed-up bunches of leaves as 'sponges' to extract water from hollows in tree stumps. Such behaviour is interesting because at one time it was fashionable to regard tool-making as the ultimate criterion of humanity. Adriaan Kortlandt has observed chimpanzees attacking with branches and stones a stuffed leopard which he placed near them. This sort of behaviour among chimpanzees has led some to conclude that tools were first used as weapons by the forebears of man.

Fewer behavioural studies have been made of the gorilla, genus *Gorilla*, but Dr George Schaller's observations of mountain gorillas have shed much light on the social organisation of this animal. The gorilla is exclusively a forest-dweller, the lowland (western) form living in parts of equatorial West Africa, and the mountain (eastern) gorilla occupying a few high areas of Central and East Africa. Spending much of their lives on the ground, gorillas are knuckle-walkers. Like the chimpanzee, the gorilla has been referred to as a 'modified brachiator', but in fact only infant gorillas ever brachiate, and then only rarely. Adult gorillas, especially the males, are too large and heavy to do more than clamber around at the lower levels of the forest. Nests are made, as by the chimpanzee and orang, but usually on the ground. Unusually for a hominoid, the gorilla is herbivorous rather than fruit-eating.

Contrary to their popular image, gorillas are gentle, easy-going animals. They live in smallish groups, of from 5 to 30 individuals, which are generally stable in composition, though individual males may wander from troop to troop. These groups range over areas of from 10 to 15 square miles, but daily travel is short, feeding being the primary occupation. The ranges of different troops may overlap, but there is aggression when troops meet. A large male acts as leader of the group, but dominance interactions are few, individuals being very tolerant of each other, as they are of other animals.

5 Human Specialisations

Man differs from his closest relatives, the apes, and from all other primates, in a multitude of ways, but those differences most evident in the skeleton, which is all we have in the fossil record, are those associated with two functional complexes: mastication and locomotion. The morphological consequences of an upright bipedal type of locomotion are evident throughout the human body, but since the evolution of structures depends upon the advantages conferred upon them by behaviour, let us first mention some of the movements involved in human locomotion. The uniqueness of human locomotion lies in its striding gait, which involves full extension of the leg at the hip and the concentration of propulsive thrust in the big toe. Most of the refinements of human walking are made possible by adaptations which maintain the centre of gravity within the smallest possible area during movement. These refinements are so successful that the vertical and lateral displacements of the centre of gravity during normal walking are contained within a 4-in. square. This is accomplished during striding by the interaction of several components of movement: horizontal displacement of the centre of gravity is reduced by the rotation of the pelvis and the bones of the leg, while its upward displacement is minimised by tilting of the pelvis, by initial flexion of the extended leg as the foot hits the ground, and by co-ordination of the movements of knee and ankle.

In man, an upright stance is effortlessly maintained by the alignment of the vertebrae of the spine in an S-shaped curve which allows the head to balance perfectly atop the neck. One conspicuous morphological correlate of this is that the vertebrae of the *lumbar* (lower back) region, and the cartilaginous discs between them, are wedge-shaped in side view, forming a forward-jutting curve. The spine attaches to the pelvis in the *sacral* region which, because of the lumbar curvature, slopes back and down. The pelvis itself is wide and shallow, and in consequence of the upright posture acts to support the abdominal viscera as well as providing areas for muscle attachment, leg articulation and, in the female, the canal for the passage of the infant at birth. The *acetabulum*, or hip-socket, is large, as is the head of the *femur* (thigh bone), which fits into it, in order to absorb the shocks transmitted from the ground during locomotion, and to accommodate the powerful muscular forces exerted during movement at the joint. Although the acetabula are wide apart because of the great breadth of the bowl-like pelvis, the femora converge towards the knees, where their distal ends are held almost together. The angle so formed is known as the *carrying angle*, and ensures that the axis of weight transmission from the feet up through the lower leg remains close to the lateral centre of gravity of the body. It also ensures that the pelvis and thigh can rotate around the axis formed by the *tibia* and *fibula* (bones of the lower leg), while the centre of gravity and direction of progression remain in a straight line.

The human foot is a highly specialised organ, adapted solely to striding bipedalism, all prehensility (capacity for grasping) having been lost. This is most clearly seen in the abduction of the big toe, which is set in line with the other toes. The big toe is also robust, since it bears most of the force exerted when the foot pushes off for each step (weight is borne on the heel as the foot strikes the ground, is transmitted along the outside of the foot as far as the ball, traverses the ball of the foot, and is finally borne mainly by the big toe). The bones of the toes (*phalanges*) are short and straight, as are those (the *metatarsals*) which lie between the toe and the ankle. In the metatarsal region the foot is sprung on two arches, longitudinal and transverse, the latter unique to man; the ankle joint is built to restrict side-to-side motion, and can move in this way only to the extent required to absorb shocks transmitted from uneven ground. The pongid foot is more mobile, and is capable of a much greater degree of prehensility, but it lacks the stability as a propulsive strut possessed by the completely terrestrial human foot.

Perhaps the best way of bringing out the efficiency of the bipedal adaptations of man is to compare human erect locomotion with that of the chimpanzee which, though it is capable of some degree of bipedalism, is adaptively a quadruped. Chimpanzees have a short spine which lacks lumbar curvature, and in consequence are unable fully to extend the thigh. When a chimpanzee is standing or moving quadrupedally it is stable because its centre of gravity falls within the area inside its four supports. But when it stands or walks bipedally the centre of gravity falls above and in front of the hip joint, whereas in man it lies directly above. In bipedal standing, the hip and knee joints are in constant flexion, and the leg muscles must work continuously to maintain the animal's position. When the chimpanzee walks bipedally, its knees are held wide apart because its femora lack a carrying angle; the legs must therefore describe wide arcs during

FIG. 5.1. SOME BIPEDAL ADAPTATIONS

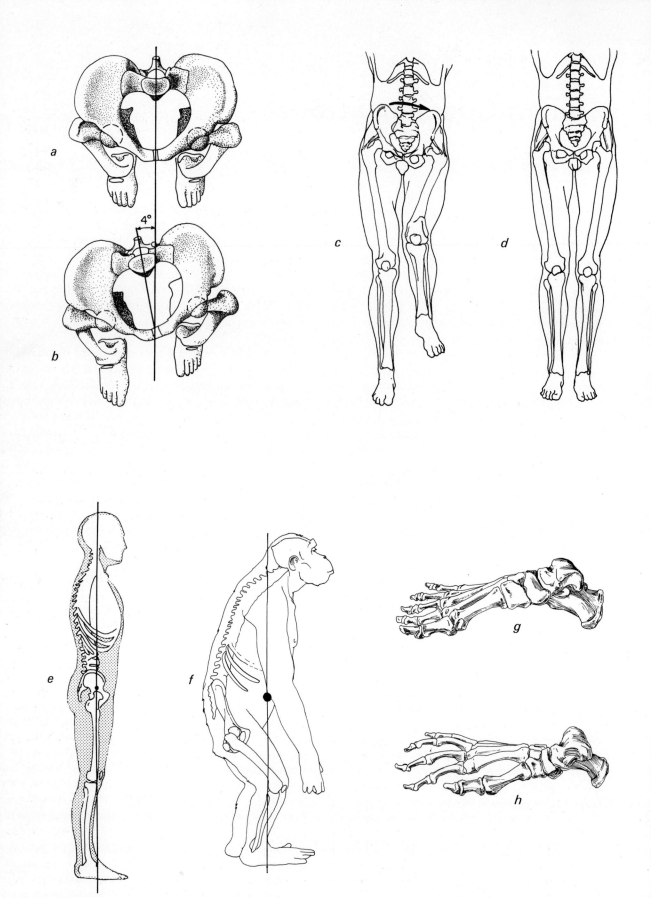

FIG. 5.1. SOME BIPEDAL ADAPTATIONS

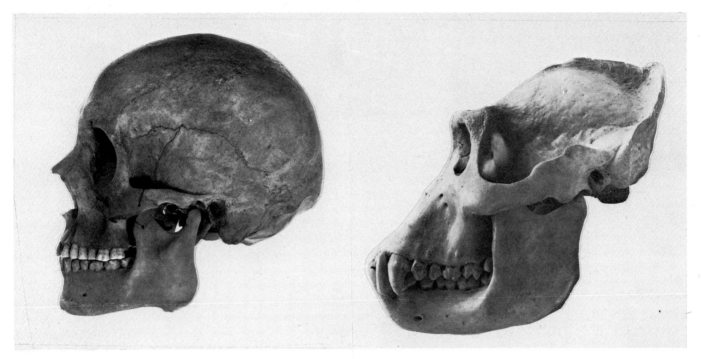

FIG. 5.2.   Comparison of human skull (*left*) with that of a male gorilla. (Not to scale.)

motion, the animal leaning from side to side as it progresses. The centre of gravity is therefore constantly shifted laterally instead of moving forward in a straight line, and a great deal of muscular effort is wasted. The chimpanzee foot is adapted for grasping; the big toe is set at an angle to the other toes, and the foot lacks a ball, so that in this area it cannot be flexed as a unit. Instability during bipedal standing results.

The reasons for which erect bipedalism was adopted in the hominid line are obscure, although, as we shall see later, hominid ancestors may in a sense have been preadapted to it. The typical hominid hunting pattern, involving fully evolved bipedalism, is very successful, but in its early stages, when bipedalism was behavioural, rather than structural, it cannot have been efficient. We must conclude, then, that the advantages of bipedalism purely as a mode of locomotion may not have been sufficient to influence its adoption by early hominids unless their behaviour was considerably different from that characterising later ones. Many non-locomotor benefits of bipedalism have been proposed as the crucial factor leading to its adoption: the ability to carry food and other objects; increased visual range; apparent increase in size; freedom of hands from locomotion for functions such as tool-using; and so forth. Of these, the last has been the most favoured among anthropologists, but, at best, it must remain speculation. In fact, probably a whole complex of advantages was involved in influencing the adoption of erect bipedalism, but at the details of this we can only guess.

The skull consists of two basic functional units: the face and the braincase. In modern man the braincase is a high, rounded vault, which provides ample area for the attachment of the temporal muscles of the jaw. Among the apes, in which the braincase is smaller, and the chewing muscles larger, a *sagittal crest* may (especially in males) be formed along the midline of the top of the skull to provide additional surface-area for the origin of the temporal muscles. Beneath the braincase there is an opening, the *foramen magnum*, through which the spinal cord descends from the brain. In man the foramen magnum is situated relatively far forward to permit the easy balancing of the skull on top of the spine, while in the apes this foramen is placed towards the back of the skull in correlation with their quadrupedal posture. The pongid skull is therefore not balanced efficiently as it is in man, the greater part of its weight lying in front of its centre

---

FIG. 5.1.   Some bipedal adaptations.

(*a*) and (*b*) illustrate the amount of pelvic rotation involved in normal walking. (*c*) shows the orientation of the pelvis during standing, while (*d*) demonstrates how the pelvis is tilted, as shown by the arrow, during striding. Note that the femora converge towards the knees. This is known as the 'carrying angle'.

(*e*) and (*f*) illustrate the centre of gravity in man and chimpanzee during standing: note how in man the centre of gravity passes directly through the hip, knee and ankle joints, while the chimpanzee is imperfectly balanced.

(*g*) shows a human foot skeleton compared with (*h*) that of a chimpanzee. The chimpanzee foot possesses a slight longitudinal arch, but not a transverse one.

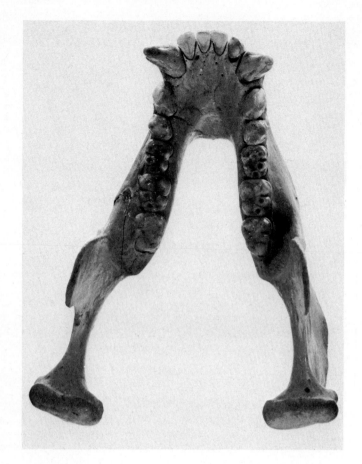

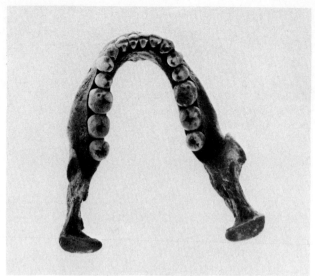

FIG. 5.3. Comparison of gorilla mandible and lower dentition with that of a human (*right*). (To scale.)

of gravity. In compensation the neck muscles of the apes are enlarged, which frequently leads to the formation of a *nuchal crest* across the width of the occipital bone. Among male primates, in which the teeth may be used in fighting, the necessity for powerful movements of the head provides additional reason for the enlargement of the neck musculature.

The shape of the facial region is largely determined by the dentition. In man, the tooth-rows, and the face, are extremely short in a front-to-back direction; the reverse is true of the apes. The incisor and canine teeth of man are small, and have become incorporated into a single slicing edge at the front of the jaw; in apes the canines, especially those of males, are robust and long, and project well beyond the general level of the tooth-row, interlocking when the mouth is closed. This affects the morphology of the premolar teeth; the front premolar (known to palaeontologists as the *third* premolar since two of these teeth have been lost in evolution) of the lower jaw, instead of being broad and possessing two cusps situated side by side, is elongated and has but a single main cusp, with perhaps a secondary one behind and inside it. This *sectorial* shape enables the long upper canine to be accommodated when the teeth are in occlusion, and may in some cases also serve as a 'whetstone' for sharpening the upper canine. The rear (fourth) premolar is bicuspid. Because the lower premolars of apes are of different shapes, they are known as *heteromorphic*, while man's, both bicuspid, are described as *homomorphic*. The molar teeth of the

gorilla are high-crowned, with high, pointed cusps and a large surface-area. Such teeth are adapted to chewing large amounts of tough, fibrous vegetable matter. Chimpanzees are frugivorous, and their molars more closely resemble the small, square, low-cusped molars of humans than do those of the gorilla.

A question which has for a long time interested anthropologists concerns the reasons for the reduction of the hominid canine. In nearly all genera of higher primates the male canine is considerably bigger than that of the female. Although male apes are known to use their canines in shredding tough vegetable material, large canines are unlikely to represent a specific feeding adaptation because there are no dietary differences between the sexes. It has been fashionable to suppose that reduction of the hominid canine has occurred because weapons replaced the function of canines in fighting—an activity which is certainly largely reserved to males. But large canines would serve as a useful adjunct to weapons during fighting, and it must be remembered that a positive evolutionary advantage should be involved in the reduction, as well as expansion or alteration, of organs. Large canines would in this view be regarded as preadaptations, rather than as adaptations, to fighting. Another suggestion is that reduction in canine size among hominids is concomitant to hormonal changes consequent upon the changed role of the male in a co-operative society. Certainly, the key may lie in subtle social changes in social organisation which we have not as yet been able to define. Or alternatively, and more probably, the positive selective advantage may lie in the incorporation of the canines into the incisor, biting, complex; Dr R. G. Every has pointed out that shortening of the face, bringing the biting teeth closer to the point where force is exerted by the muscles of the jaw, greatly increases the cutting power of the human anterior tooth complex.

6 Palaeocene and Eocene Primates

Since teeth, with their durable enamel coating, are the parts of animals most frequently preserved as fossils, our knowledge of early primates is largely confined to the dentition. Primates appear to have evolved from primitive 'insectivores' towards the end of the Cretaceous, probably as a result of a change among the ancestral primates from an insectivorous to a frugivorous diet, which of course wrought profound changes in the masticatory and digestive systems of the animals concerned. The teeth of early primates, in contrast to those of other early mammals, have lower, blunter cusps, and were better suited to chewing than were those of contemporaneous insectivores. A frugivorous diet, involving the handling of food, may also have provided an impetus towards increased intelligence in the primate line, and be correlated with a change from a nocturnal to a diurnal activity rhythm.

The earliest possible primate is *Purgatorius ceratops*, known by a single tooth from the late Cretaceous of North America, about 70 million years ago, though this tooth is of archaic aspect and might equally well represent an ancestral condylarth (archaic ungulate). The Palaeocene witnessed a radiation of forms which are regarded as primates primarily because of resemblances in their cheek teeth to those of later primates; it has not so far been possible, however, to discern any specific ancestor-descendant relationships between the primates of the Palaeocene and those of the succeeding Eocene epoch, and it must be assumed that all presently known Palaeocene forms usually regarded as primates became extinct without issue.

The best-known Palaeocene primate is *Plesiadapis*, which occurs in the late Palaeocene of both France and North America. *Plesiadapis* was long-snouted, and undoubtedly possessed a well-developed sense of smell; it lacked a post-orbital bar (a strut of bone separating the orbit from the temporal fossa—an extremely primitive condition among primates); and its eyes faced sideways. Its cheek teeth were

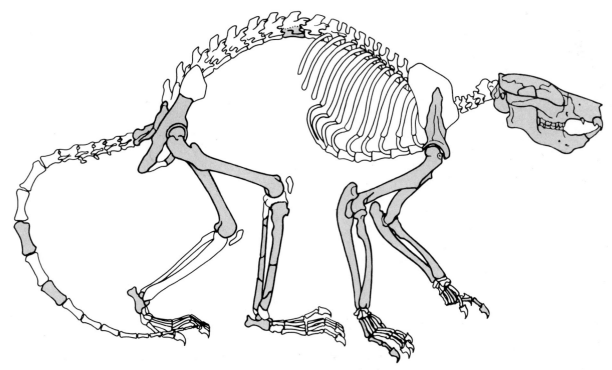

FIG. 6.1. Skeleton of *Plesiadapis tricuspidens* from the late Palaeocene of France. Shaded areas represent known elements. (Approximately one-half natural size.)

25

very similar to those of *Pelycodus*, an undoubted primate of the Eocene, but its anterior teeth were peculiarly specialised and rather like those of rodents. Postcranially, *Plesiadapis* bore similarities to modern squirrels, but was extremely heavily built—perhaps too heavily built to have been very efficiently arboreal. Its fore- and hind limbs appear to have been of approximately equal length, and its forelimb was adapted to very strong flexion—a characteristic of vertical clingers and leapers. Perhaps the locomotion of *Plesiadapis* was preadapted to the vertical clinging and leaping which Dr Alan Walker has shown to have been characteristic of most, if not all, Eocene primates. However, *Plesiadapis* still possessed claws on all five digits, and probably did not have grasping extremities.

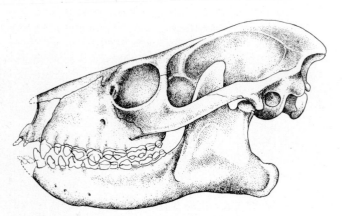

FIG. 6.2. Skull of *Notharctus*, a North American Eocene primate. (Natural size.)

The differences between Palaeocene and Eocene primates are radical. Some remnants of the Palaeocene radiation of primates survived into the Eocene, but this epoch was essentially the time of radiation of an entirely new kind of primate, still only prosimian in grade, but a great deal more advanced than its predecessors and with clearly recognisable affinities to the prosimians of today. Although such prosimians clearly flourished also in the forests of Eocene Africa, their remains are only known from North America and Eurasia. The emergence of these more advanced primates has been suggested to corr:late with the radiation at the end of the Palaeocene of the rodents, competition with which, according to this theory, forced the primates out of their old niche and into a new, highly arboreal one.

Eocene primates had reduced snouts, indicating a reduction in importance of olfaction; correlating with this, their eyes were forward-facing, providing overlapping fields of vision and thus probably some degree of stereoscopic sight. These adaptations are both concomitant to an arboreal way of life in which sense of smell is relatively unimportant, while the ability to judge distances is vital to survival. Their locomotor habit, as we have seen, was totally arboreal; perhaps the most profound morphological change associated with this was the acquisition of a grasping hand, in which the first digit is opposed to the others. Claws were lost, to be replaced by flat nails supporting tactile pads at the ends of the fingers. By the Eocene, primates already possessed brains which were relatively larger than those of contemporaneous non-primates; and in contrast to their Palaeocene ancestors, which were probably solitary, these animals almost certainly lived in social groups.

It is generally difficult to define the evolutionary relationships between Eocene and later primates, but it seems likely that the Eocene subfamily Microchoeriinae contained the ancestor of modern *Tarsius*, while a member of the family Adapidae may have given rise to the lorisoids, and a related African form, of course unknown, could have been the progenitor of the modern Malagasy lemuroids. The Old World higher primates appear to have originated in Africa, although this cannot be substantiated at present, and a plausible, if hypothetical, ancestor could have been an African member of the family Omomyidae, known from the Eocene of both North America and Europe.

7 Oligocene Primates

The Old World and New World higher primates evolved separately from prosimian stocks at some time during the late Eocene, or, more probably, during the early Oligocene. Despite these separate origins, both groups share a number of features distinguishing them from prosimians. Among these are: increase in relative brain size; formation of a complete bony wall behind the orbit; further improvements in vision and orbital frontality, and increased reduction in the sense of smell; faces tucked in beneath the braincase; teeth of simpler morphology; a change in jaw mechanics with the raising of the jaw articulation; and behavioural, manipulative and postcranial changes.

Apart from a few surviving North American prosimians, our entire knowledge of Oligocene primates derives from a small area of the Egyptian desert known as the Fayum. Today the Fayum is an area of desert badlands, but in the Oligocene it was covered by lush tropical forests and crossed by numerous watercourses. In this forest primates abounded. Until a few years ago, the entire record of Fayum primates consisted of a handful of fragmentary specimens which nevertheless formed the basis for a great deal of speculation upon the origins of higher primates. In 1961, however, Dr Elwyn Simons, of Yale University, began a series of fossil-hunting expeditions to the Fayum. His discovery of new forms, and of additional material of species already known, has corrected a large number of misconceptions, and has paved the way for a fuller understanding of higher primate origins and relationships.

The Fayum deposits consist of two formations, the Qasr el-Sagha beneath, composed of Eocene marine deposits, and the terrestrial Jebel el-Qatrani above, which is capped by a basalt flow dated by K-Ar at about 26 million years. In places, the Jebel el-Qatrani was almost completely eroded away before deposition of the basalt, so the top of this formation may be a million years or more older than the basalt. The Jebel el-Qatrani, in which all the Fayum fossils have been found, is composed of four sections (Fig. 7.1). Nearly all the primate fossils have been found in two 'fossil wood zones' which contain large fossilised tree trunks; and only four quarries, E, G, I and M, have yielded primates at all.

Most Fayum primates belong to one of two families: Parapithecidae and Pongidae. The parapithecids are the most abundant Fayum primates, and fall into two genera: *Apidium* and *Parapithecus*, which are unique among Old

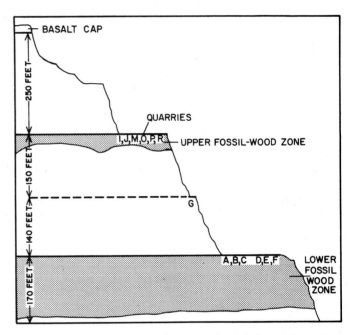

FIG. 7.1. Section through the Oligocene Jebel el-Qatrani Formation, Fayum, Egypt.

World higher primates in possessing three premolar teeth on each side of the jaw, above and below, and both of which are of similar morphology; they are quite obviously closely related forms. *Parapithecus*, however, does differ from

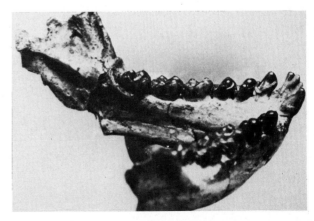

FIG. 7.2. Mandible of *Parapithecus fraasi*. (Approximately twice natural size.)

Courtesy of Dr E. L. Simons

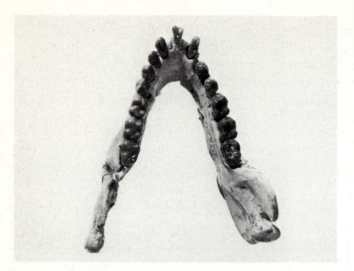

FIG. 7.3. Mandible of *Apidium phiomense*. (1·5 times natural size.)

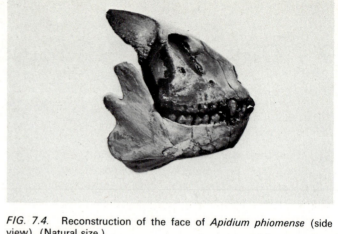

FIG. 7.4. Reconstruction of the face of *Apidium phiomense* (side view). (Natural size.)

Courtesy of Dr E. L. Simons

Apidium in possessing molars which are more 'waisted'; i.e. at their midpoint they are very distinctly constricted from side to side. This is likewise a feature of the cercopithecoid monkeys, and it is possible that in *Parapithecus* we have a cercopithecoid ancestor. In fact, some authorities would place Parapithecidae as a subfamily of Cercopithecidae, Parapithecinae, on the strength of this possibility. Another difference between the two genera is that *Apidium* possesses a centroconid cusp in the middle of its lower molars, a feature which it shares with the strange Pliocene primate *Oreopithecus*, and which has been suggested to indicate a phyletic relationship between the two forms. At present, however, this must remain no more than a suggestion. The face of *Apidium* was short, which correlates with the reduction in the olfactory lobes evident from a frontal bone which we have of this genus. The frontal also shows that the orbit was entirely encased in bone behind, a typical feature of higher primates. Preliminary examination of some post-cranial bones, however, suggests that the parapithecids had a leaping form of locomotion not unlike that of some modern Malagasy lemurs.

Two Fayum genera are assigned to Pongidae: *Propliopithecus* and *Aegyptopithecus*. The type specimen of *Propliopithecus*, a lower jaw, was discovered early in this century by a German fossil hunter, Richard Markgraf, and was named by his compatriot Max Schlosser in the erroneous belief that it represented an antecedent of the Pliocene gibbon ancestor, *Pliopithecus*. *Aegyptopithecus* is a much more recent discovery, the first remains of this animal having come to light in Quarry I in 1964. It was only in 1966, however, that a complete skull of this animal was recovered, from the same stratigraphic level as the type, at Quarry M. The age of this level is probably around 29 million years. The skull of *Aegyptopithecus* is extremely primitive. It has a long snout, although the olfactory bulbs of the brain are relatively reduced, and the eyes face forward. The snout is largely composed of the pre-maxillary bones,

which are reduced in modern apes, and entirely absent in man. The orbits are encased behind by bone, but the ear region is primitive, resembling that of the lorises in having a fused tympanic ring; there is no bony tube leading from the eardrum to the outside of the skull. Muscle markings on the temporal bones suggest that *Aegyptopithecus* possessed a small sagittal crest—a suggestion reinforced by the large size of the temporal fossa, the area in front of the ear which accommodates the jaw muscles.

Dentally, *Aegyptopithecus* unquestionably lies in Dryopithecinae, the subfamily of early apes, and in fact bears some remarkable similarities to *Dryopithecus africanus*, an ancestral chimpanzee of the Miocene. *Aegyptopithecus* has large canine teeth, and sectorial anterior lower premolars; the upper molars possess large, beaded internal cingula (enamel ridges running along the perimeter of the tooth; a primitive characteristic among apes) and the lower molars large external cingula. The lower molars increase in length from front to back (likewise a primitive ape characteristic). The total dental pattern of *Aegyptopithecus* is thus unequivocally pongid, albeit primitive.

It is interesting that *Propliopithecus* differs dentally from *Aegyptopithecus* in many of the ways in which hominids differ from pongids. The lower molars of *Propliopithecus* show no increase in size from front to back; its canines are diminutive (easily explicable if the type specimen, the only one known which possesses canines, represents a female); its lower premolars are less heteromorphic; and the horizontal rami of its mandible are broad and shallow, rather than narrow and deep, as in *Aegyptopithecus* and other apes. However, a more reasonable explanation of these differences than that Hominidae and Pongidae were already distinct by the late Oligocene is that *Propliopithecus* was a generalised, short-faced primate which had not yet acquired many of the specialisations associated with later hominoids. The locality at which the type of *Propliopithecus* was recovered is unknown, but it seems likely that it came from the

FIG. 7.5.
Basal and lateral views of the cranium of
Aegyptopithecus zeuxis. (Natural size.)
Courtesy of Dr E. L. Simons

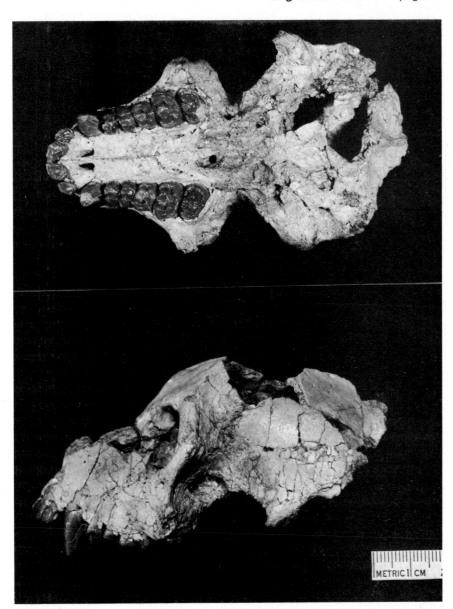

FIG. 7.6 (below left). Lateral and occlusal
views of the mandible of *Aegyptopithecus
zeuxis.* (Natural size.)
Courtesy of Dr E. L. Simons

FIG. 7.7 (below right). Type mandible of
Propliopithecus haeckeli. (1·1 times natural
size.)

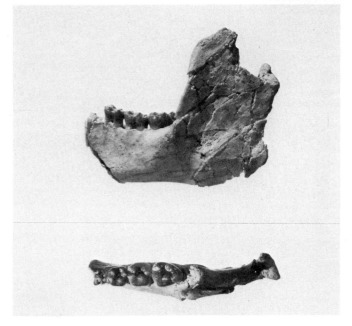

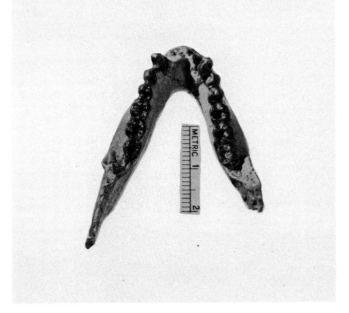

lower fossil wood zone, and is therefore perhaps 32 million years old—older than any *Aegyptopithecus* yet found. In view of the detailed similarities in cheek-tooth morphology between *Aegyptopithecus* and *Propliopithecus*, it seems possible that the latter represents the ancestral type from which *Aegyptopithecus* arose, especially since a series of isolated teeth from Quarry G, as yet undescribed, show resemblances to both forms and may plausibly be regarded as intermediate.

Thus, in the absence of any firm evidence to the contrary, it appears that in *Aegyptopithecus*, a pongid, we may have the progenitor of both the hominid and pongid lines. The time of separation of these lines, however, is still uncertain; a variety of early Miocene pongids is known, all of which seem too specialised for the role of ancestral hominids, but hominids are not known until the very end of the Miocene, some 14 million years ago.

Finally, two little-known Fayum primates can be regarded as of uncertain familial affinities. One of these is *Oligopithecus savagei*, which comes from Quarry E, lower in the section than any other Fayum primate. *Oligopithecus* is only known from a jaw fragment containing a few teeth, but this is enough to tell us that unlike the parapithecids it possesses only two premolars. Characteristics of the mandible likewise proclaim its affinities as being with the Old World higher primates, although its molar teeth are most reminiscent of those of the Eocene omomyids. *Oligopithecus* is best regarded as being the most primitive known catarrhine primate.

The other problematical Fayum primate is *Aeolopithecus chirobates*, known from a single jaw found in Quarry I in 1963. All teeth are present in the jaw except the incisors, but chemical weathering has destroyed their cusp patterns. *Aeolopithecus* is small, but its canines are relatively enormous, and its anterior premolars large. It has been suggested that the decrease in the depth of its jaw towards the rear, the form of its symphyseal cross-section and the reduction of its third molars may indicate affinities with the gibbons, but until better remains are known this cannot be demonstrated with any certainty.

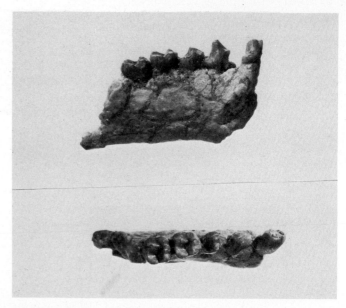

FIG. 7.8.   Internal and occlusal views of type specimen of *Oligopithecus savagei*. (Twice natural size.)

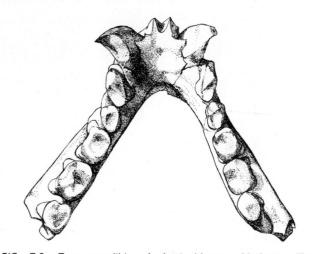

FIG. 7.9.   Type mandible of *Aeolopithecus chirobates*. (Twice natural size.)

Courtesy of Dr E. L. Simons

8　The Rise of the Pongids

The first discovery of a fossil ape was made in 1856, when Edouard Lartet described a mandible from middle Miocene deposits at St Gaudens, in France. This he named *Dryopithecus fonani*. Since that time, closely related forms of Mio-Pliocene age have been recovered from Africa and Asia, as well as from Europe. Although most of these fossils were assigned to the subfamily Dryopithecinae, there was an enormous proliferation of names at the generic and specific levels. This explosion in nomenclature was due primarily to failures by palaeontologists to realise the extent of morphological variation which occurs in modern pongid species and genera and also, presumably, occurred in their ancestors; plus the evolutionary changes between forms sampled at different times from the same lineage. Students confronted with over 25 proposed genera and over 50 proposed dryopithecine species have tended to assume that pongid evolution during the Miocene and Pliocene consisted of a very complex adaptive radiation. Recent work by Simons and by Dr David Pilbeam has shown that this was not the case. Nearly every Mio-Pliocene pongid fossil may be assigned to the genus *Dryopithecus*, while some proposed dryopithecines do not belong in this subfamily, having proved instead to be monkeys, hylobatids or hominids.

In 1933 A. T. Hopwood, of the British Museum (Natural History), described a series of middle Miocene pongid fossils from Kenya, naming them *Proconsul africanus*, and suggesting that they might bear an ancestral relationship to the chimpanzee. Since then, a large number of fossils of this genus have been found in Kenya and Uganda. In 1950, it was shown that the material studied by Hopwood actually consisted of four species: *P. africanus*. *P. nyanzae*, *P. major*, and *Sivapithecus africanus*, of which the status is now uncertain. The work of Simons and Pilbeam indicates that these all belong in genus *Dryopithecus*. However, they are separable at the subgeneric level from other variants of *Dryopithecus*, and their former generic names are therefore placed in parentheses between the generic and trivial names. For instance: *Dryopithecus (Proconsul) major*. The only site from which all of these are definitely known is the island of Rusinga in Lake Victoria, whose fossiliferous deposits are about 18–19 million years old. Another site in Kenya, Songhor, has yielded (*S.*) *africanus*, (*P.*) *africanus*, (*P.*) *major* and possibly (*P.*) *nyanzae*. The site at Koru, close to Songhor, has produced (*P.*) *africanus* and possible representatives of the other two (*Proconsul*) species; both former

sites are dated at around 19·5 million, and the latter is probably of the same age.

D. (P.) africanus is small, and probably ancestral to the chimpanzee. Morphologically its molar teeth, apart from

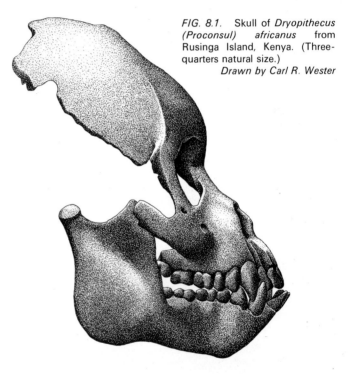

FIG. 8.1.　Skull of *Dryopithecus (Proconsul) africanus* from Rusinga Island, Kenya. (Three-quarters natural size.)
Drawn by Carl R. Wester

bearing strong cingula, are very similar to those of *Pan*, and significantly resemble those of *Aegyptopithecus*. Chimpanzees have huge expanded incisors, a correlate of their frugivorous diet, and the slight expansion of those of (*Proconsul*) *africanus* probably indicates that the animal was on its way to becoming frugivorous. A skull of (*P.*) *africanus* was discovered in 1948 on Rusinga by Dr L. S. B. Leakey. Since the teeth of (*P.*) *africanus* are small, and the brain is relatively expanded, there is no cresting and the skull vault is delicate and rounded. Also because of the small dentition, the face is short and gracile, and the mandibular symphysis (where the left and right halves of the lower jaw meet) lacks a simian shelf (a buttress of bone which braces the bottom of the inside surface of the symphysis, resisting torsion, and assisting in the transfer of chewing stresses across the jaw). Postcranially, (*P.*) *africanus*, the size of a small female baboon, appears to have been an active quadruped in some ways

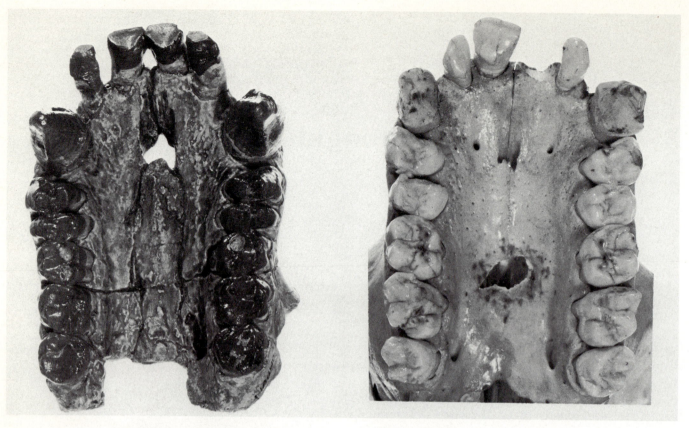

FIG. 8.2. Comparison of the palate and upper dentition of *Dryopithecus major* from Moroto, Uganda (*left*), with that of a modern female gorilla. (Both natural size.)

generalised, yet already showing some adaptations to quadrupedal knuckle-walking.

D. (P.) nyanzae is a slightly larger species which probably represents a terminal development. Pilbeam has suggested, however, that it may resemble the common ancestor of the other two (*Proconsul*) species.

D. major, males of which were the size of large chimpanzees, bears a number of striking resemblances to the gorilla, to which it is almost certainly ancestral. The sites from which it is best known appear to indicate an ecology similar to that of the gorilla: thick forest on the slopes of volcanoes. Among gorillas there is a great deal of sexual dimorphism, especially in the size of the body and of the canine teeth. For a long time the only fossils ascribed to *D. major* in which these teeth were present had large canines. Pilbeam's researches showed, however, that a number of small-canined fossils placed in *D. nyanzae* were better viewed as females of *D. major*, a strongly dimorphic species. Postcranial remains of *D. major* are sparse, but it seems that this species may have been more terrestrial than *D. africanus*.

As was mentioned at the beginning of this chapter, the first dryopithecine was found in Europe. Since then, further pongid material has been recovered from the Miocene and Pliocene of this area. These European dryopithecines became extinct without issue, gradually dying out as the climate of Europe became increasingly inhospitable.

Further to the east, in the Himalayan foothills of North India and West Pakistan, a large number of *Dryopithecus* fossils has been found. As happened elsewhere, many names were proposed for these fossils, but in the recent revision by Simons and Pilbeam, they were reduced to two: *D. (Sivapithecus) sivalensis*, a smaller species, and the larger *D. (S.) indicus*. It now appears that there may just possibly be only one species represented, by priority *D. sivalensis*, and that the Indian and European dryopithecines might better be placed in the single subgenus *D. (Dryopithecus)*. It is tempting to infer an ancestor-descendant relationship between *D. sivalensis* and the only modern Asian pongid, *Pongo*, and, indeed, some similarities between them can be found. However, such a relationship must remain no more than a possibility, based primarily on the fact that *D. sivalensis* at present remains the only candidate for orang ancestry.

One of the most interesting primates to be found in India was recovered in 1968 by an expedition under Dr Simons's direction. This middle Pliocene pongid, perhaps 7 or 8 million years old, was named *Gigantopithecus bilaspurensis*, and is antecedent to *G. blacki* of the Chinese Pleistocene, probably the largest primate ever to have lived. *G. bilaspurensis* is known from a single lower jaw, with all teeth represented except the incisors which must, however, have been small, since there is very little space between the canines. These latter teeth appear to have become incorporated functionally in the battery of grinding teeth (premolars and molars), which have broad, flat occlusal surfaces, and are closely packed together. The mandible is very massive and powerful. An early Pliocene molar tooth intermediate in morphology between those of *G. bilaspurensis* and of *D. indicus* strongly suggests that *Gigantopithecus* was derived from *D. indicus*, perhaps towards the end of the Miocene.

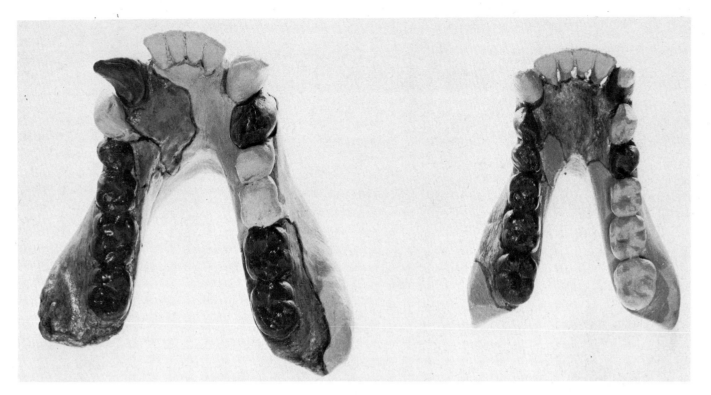

FIG. 8.3. Reconstruction of the mandibles of *Dryopithecus indicus* (*left*) and *D. sivalensis* from the Siwalik Hills of India.

E. L. Simons reconstruction

G. blacki is known from three mandibles and numerous isolated teeth from China, perhaps 0·5 or 0·75 million years old. Males appear to have been much larger than females, with rather bigger canines. *G. blacki* is broadly similar to *G. bilaspurensis*, though it is larger, with higher-crowned and more cuspidate cheek teeth, and with more distinctly bicuspid anterior lower premolars.

From the size of its jaws, we may infer that *Gigantopithecus* was at least as large as, and probably larger than, the gorilla; it was therefore almost certainly a terrestrial form. Such an impression is reinforced by the fact that the earlier species was found in association with an open-country fauna, and there is no reason to suppose that its descendant did not occupy a similar habitat. It seems not too unlikely, then, that *Gigantopithecus* foraged for food in a wooded savanna, or even a more open setting.

Recent work by Dr Clifford Jolly has shown the relationship between the diet and feeding methods of the gelada baboon and some of the features of its skull and dentition. Geladas feed, as we have already seen, by putting small morsels into their mouths. Their incisors are therefore reduced, and their grinding batteries accentuated. Continuous heavy use of their cheek teeth causes crowding and rapid wear, a feature also seen in *Gigantopithecus*. The dental mechanisms of the gelada and *Gigantopithecus* are similar enough to allow us to infer that they possessed similar diets and feeding habits. In several features of its dentition and mandible, *Gigantopithecus* parallels early hominids, and several authorities have suggested that it should be placed in Hominidae. However, it is clear that *Gigantopithecus* is a pongid, albeit a highly specialised one.

Before ending this chapter we should mention two Mio-Pliocene primates which, although not pongids, deserve discussion: one because it is ancestral to the gibbon, the other because of its total uniqueness.

In East African deposits, together representing almost the whole of the Miocene from about 22 to about 14 million years ago, have been found the remains of a creature known as *Limnopithecus*. The skull and teeth of this animal strongly resemble those of the gibbon, to which it is almost certainly ancestral, but its postcranial skeleton is much more generalised than the gibbon's, resembling in some respects that of *Ateles*. Its arms and legs are of approximately equal length, and, again unlike the gibbon, it possessed a tail. We need not be surprised at this; mosaic evolution is a well-documented phenomenon, and if evolution has occurred, ancestors can hardly be expected to look like their descendants! But certainly, *Limnopithecus* must have been doing some arm-swinging. By the Pliocene, hylobatids had reached Europe, and various sites of Pliocene age have yielded the remains of *Pliopithecus*, a form descended from *Limnopithecus* which may not, in fact, have evolved sufficiently to warrant separate generic status. Also by the Pliocene, hylobatids had disappeared from Africa, probably because of competition from other primates, which may also account for the adoption of brachiation as a means of feeding in the most peripheral areas of the trees, the only part of the canopy unavailable to the quadrupedal cercopithecoids.

In swamp deposits in Tuscany, dating from the end of the Miocene or the beginning of the Pliocene, have been discovered the remains of one of the most enigmatic primates ever found. The cheek teeth of this animal, *Oreopithecus*, are quite unlike those of any other primate, although the fact that its lower molars do possess a centroconid cusp has led some authorities to infer a phyletic relationship between

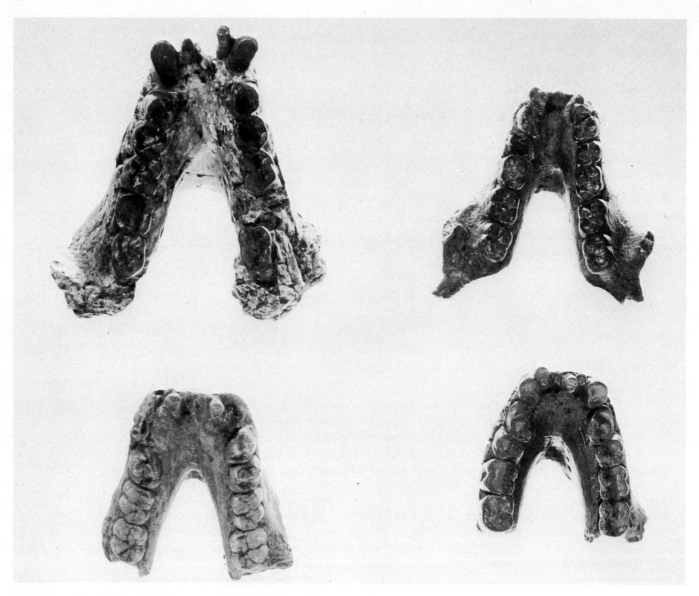

FIG. 8.4. Comparison of the type mandible of *Gigantopithecus bilaspurensis* (*upper right*) with the three mandibles of *G. blacki.* (One-half natural size.)

Oreopithecus and *Apidium.* The skull of *Oreopithecus* is rounded and globular, its face flat, and its canines small. These features have prompted the suggestion that *Oreopithecus* is hominid, but the similarities to Hominidae are certainly merely parallel functional developments. Postcranially, *Oreopithecus* is equally puzzling. It has the broad, shallow thorax and reduced lumbar region of primates with an habitual erect posture, and in its pelvis it has certain features characteristic of modern man. But its knee and ankle joints, as well as those of its arms and shoulder, are built for mobility rather than stability. Its locomotor pattern probably resembled most closely that of the orang, but this is a problem which will bear considerable further investigation. Taxonomically, *Oreopithecus* is best placed in its own family, probably within Hominoidea.

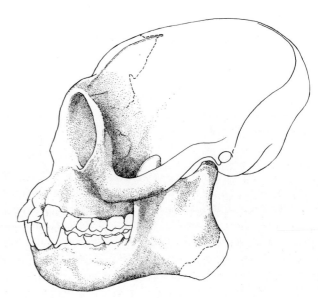

FIG. 8.5. Skull of *Pliopithecus* from the Pliocene of Europe. (Four-fifths natural size.)

9 The Earliest Hominid

We must now turn our attention back to northern India, where in the first quarter of this century, Guy Pilgrim discovered two mandibular fragments and a right maxilla (upper jaw). These specimens, of probable late Miocene or

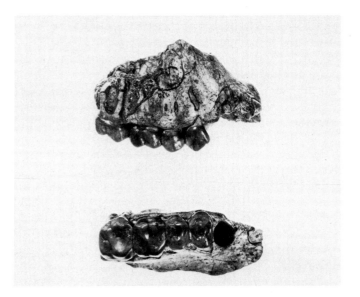

FIG. 9.1. Type maxilla of *Ramapithecus 'brevirostris'* from the early Pliocene of India. Lateral and occlusal views. (Natural size.)

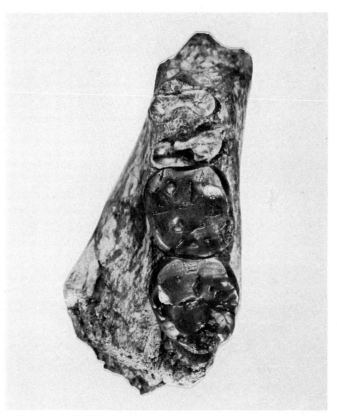

FIG. 9.2. Partial *Ramapithecus* mandible from the early Pliocene of India. (One and two-thirds natural size.)

early Pliocene age, he assigned to a new species of *Dryopithecus*, *D. punjabicus*. Later, in 1934, G. E. Lewis placed the mandibular specimens in a new genus, *Bramapithecus*. At the same time Lewis described a new primate genus and species, *Ramapithecus brevirostris*, of which the type specimen, a right maxillary fragment, is probably from the early Pliocene, about 12 million years ago. In his initial description of this form, Lewis noted many similarities to Hominidae, but it was only in his Ph.D. dissertation, never published, that Lewis stated his belief that *Ramapithecus* was ancestral to later hominids. Thus it was not until 1961, when Dr Simons restudied the specimen, that *Ramapithecus* began to gain general acceptance as an early hominid. Subsequently, it was realised that *Bramapithecus* and *D. punjabicus* also belong in *Ramapithecus*. Because *Dryopithecus punjabicus* was the first name proposed for a member of the species, but the generic name *Dryopithecus* was preoccupied, the correct name is *R. punjabicus*.

In 1962, Dr L. S. B. Leakey published a description of a new primate from a late Miocene site at Fort Ternan, in Kenya, K-Ar dated at 14 million years. For this specimen he proposed the new genus and species name *Kenyapithecus wickeri*. The *Kenyapithecus* maxilla is, however, sufficiently similar to those of *Ramapithecus* to belong in the same genus, and very probably in the same species, despite the fact that the Fort Ternan specimens are older than most if not all of the Indian material. An upper canine tooth was found close to the Fort Ternan maxilla, and may well have belonged to the same individual.

The cheek teeth of *Ramapithecus*, in contrast to those of contemporaneous apes, are steep-sided, with broad, relatively flat, squarish occlusal surfaces. They are closely packed and show a marked gradient of increasing wear from back to front. The implication of this is either that a more abrasive diet or more powerful chewing than that of the apes promoted faster wear, or that, as in man, the eruption

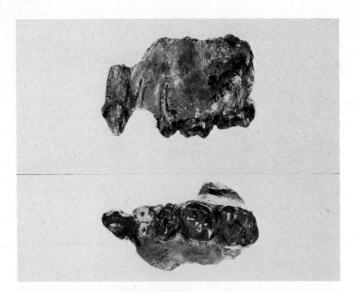

FIG. 9.3. Ramapithecus maxillary fragment (type of *Kenyapithecus wickeri*) from the late Miocene of Fort Ternan, Kenya. Lateral and occlusal views. (Natural size.)

sequence of the molars was extended over a longer period, or a combination of these. The premolars are homomorphic, while the canine is relatively reduced (the significance of which we have already discussed), even compared to the smallest female ape. Nevertheless, the canine, though small, is morphologically reminiscent of those of contemporaneous apes. This, however, is hardly unexpected if *Ramapithecus* was derived from a species best classified as pongid. The incisors are also small, and, as what is preserved of their sockets shows, vertically emplaced, while the dentition as a

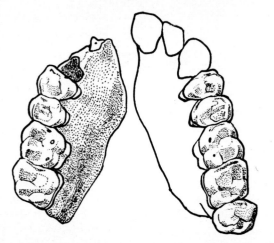

FIG. 9.4. The two maxillary fragments of *Ramapithecus* from India, positioned in occlusal view to indicate the probable shape of the dental arcade. (Slightly over natural size.)

whole was probably set in an arcade reminiscent in shape of those of early Pleistocene hominids, though not completely in the continuously curving arcade of modern man. In pongids, the large canines stand out as 'cornerstones' of a U-shaped arcade.

The evidence of the maxilla shows that the face of *Rama-*

pithecus was much flatter than that of any pongid; the mandible likewise shows this, curving sharply inwards towards the front at the level of the first molar.

The dental, facial and mandibular resemblances of *Ramapithecus* to later hominids are striking; what do they mean in terms of function? Dr Pilbeam has recently completed a functional analysis of known *Ramapithecus* specimens, and the following discussion is based on his work. The zygomatic arch, which carries the origin of the powerful masseter muscle, extends further forward on the maxilla than in pongids; this allows the masseter, which runs from the base of the ascending ramus of the jaw to the zygomatic arch, to originate more anteriorly, and therefore to have a greater power arm around the jaw joint, which provides for more powerful chewing. Since the zygomatic arch also serves to dissipate the stresses set up in the face during chewing, its origin above the first molar, and therefore above the midpoint of the cheek-tooth row, may show that all these grinding teeth were of equal importance in chewing. The shortness of the face of *Ramapithecus* allowed a greater biting force to be exerted at the front of the jaw, since the load arm of the jaw was decreased. There is also evidence to suggest that the canine and incisor teeth were in the process of becoming an integrated slicing unit; it is certainly logical that these two adaptations should be found in conjunction.

That *Ramapithecus* was relatively deep-faced is shown by the distance between the bottom of the nasal aperture and the margin of the upper anterior tooth-row; this is also suggested by certain features of the lower jaw. At the symphysis, if the specimen which shows this has been correctly identified, the jaw is relatively much deeper than in the chimpanzee, and it is restricted in width from front to back. The symphysis bears a simian shelf. The possession of a deep face and of a strongly buttressed jaw are almost certainly associated with the resolution of the strong stresses set up in these regions during powerful mastication.

The teeth of both upper and lower jaws in *Ramapithecus* had expanded occlusal surfaces and were closely packed. This tooth-packing is probably a result of 'mesial drift', caused by forward stresses along the tooth-row resulting from strong masticatory pressures.

Although still primitive then, *Ramapithecus* essentially possessed the dental mechanism seen in the early Pleistocene hominids of the genus *Australopithecus*, of which it is an extremely plausible ancestor. The deep, short face, robust mandible, and grinding posterior dentition all suggest a dental mechanism distinctly more powerful and better adapted to grinding than that of apes.

Early Pleistocene hominids are distinguished from pongids in two main ways: cranially in possessing the characteristic hominid dental/facial mechanism, and postcranially in being upright bipeds. We have seen that dentally *Ramapithecus* is unequivocally hominid. What about its locomotor mechanism? Indian *Ramapithecus* occurs almost exclusively in the Nagri formation of the Siwalik series. It has been recently pointed out that the Nagri fauna implies warm, moist conditions supporting tropical forests lining broad, sluggish rivers. There appears to have been a drying tendency throughout Nagri times, and towards the end of the Nagri there was probably local replacement of forests by open tree-savanna. We have no direct knowledge at all of the postcranial structure of the *Ramapithecus*, but in view of its environment, it was almost certainly arboreal to some extent. Probably *Ramapithecus* was still primarily arboreal in its morphology and behaviour, but was turning, as its dentition indicates, to a foraging way of life on the forest floor, at forest fringes, and in open woodland. It is probably in this last setting that hominid bipedalism evolved. Why was erect bipedalism adopted? The closest Miocene and early Pliocene relatives of *Ramapithecus* appear to have been postcranially not greatly unlike *Ateles*: basically quadrupedal, but possessing some adaptations associated with arm-swinging and, most importantly, truncal erectness. If *Ramapithecus*, or its ancestors, were adapted in this way, it would have been to some extent preadapted to a bipedal form of locomotion away from the forest. As

Pilbeam has pointed out, if the ancestors of *Ramapithecus* had been quadrupeds of the general cercopithecoid type, they would probably have evolved on the ground into digitigrade quadrupeds similar to the patas monkey or gelada baboon. Dr R. F. Ewer has observed that bipedalism is an impractical mode of locomotion in hilly, forested conditions, and it seems probable that, had *Ramapithecus* come to the ground on such terrain, it would have become a knuckle-walker like the gorilla, which occupies territory of this kind today.

If the ancestors of Hominidae are best classified as pongids, where should we draw the boundary between the two families? It has been argued that, since Hominidae is basically characterised by two major adaptive shifts, one in the dental/facial complex and one in the locomotor mechanism, any form which is ancestral to, or close to the ancestry of, later hominids and which has crossed at least one of the adaptive boundaries is properly classifiable as hominid. In this view, the status of *Ramapithecus* is unequivocal.

10   The Pleistocene

Early hominids are best known from the Pleistocene epoch, commonly known as the 'Ice Age', although widespread glaciation of the major continents occupied only its latter part. The duration of this epoch has been the subject of warm debate, concerned with the problem of when the Pleistocene actually started. Traditionally there have been two approaches to the problem. One claims that the base of the Pleistocene should be defined on the basis of the evolutionary states of fossil faunas. This approach was until recently hampered by the fact that faunal transitions are gradual, and it is largely a matter of taste as to which organisms are taken as definitive. The other approach is to define the Plio-Pleistocene boundary on the basis of glacial phenomena. Unhappily glaciations are not confined to the Pleistocene, evidence for glaciations of the most recent series occurring as far back as the Miocene. All in all, if the nineteenth-century geologists who established two epochs between the Miocene and the Recent had possessed today's knowledge, they might not have cared to make the distinction between the Pliocene and the Pleistocene.

However, if the distinction is to be made, there is an arbitrary way of doing it based on the fact that, by geological convention, each epoch must be represented by a type section, the age of the base of which determines the age of the base of the epoch. The Calabrian stage represented by the marine section at Le Castella, Italy, is the first stage of the Pleistocene epoch, and its base represents the beginning of the Pleistocene.

It is now known that the Earth's magnetic field reverses its direction at irregular intervals; such changes are detectable in the geological record. An almost complete Pleistocene Atlantic marine stratigraphic record is known from deep-sea cores, and the sequence of palaeomagnetic reversals during the Pleistocene has largely been worked out. These reversals are also detectable in terrestrial deposits which have been dated by K-Ar. In the Atlantic record, the evolutionary change of the species of planktonic foraminifera *Globorotalia tosaensis* into *G. truncatulinoides* has been shown to have taken place either just before, or during, the 'Olduvai event', a brief period of normal polarity dated by K-Ar at about 1·85 million years ago. This evolutionary transition also took place at the base of the Calabrian, so we may date the base of the Pleistocene at 1·85–1·9 million years ago. It has further been suggested that the marked increase in abundance of the foram *G. inflata*, a temperate species, at the time of the Jaramillo magnetic event, might place the onset of major continental glaciation at about 0·8 million years ago. Raimondo Selli, extrapolating from sedimentation rates in the Po plain of Italy, has arrived at similar conclusions, estimating the base of the Pleistocene at 1·8 million years and the beginning of what he calls the 'Ice epoch' at about 0·8 million years. The extraordinarily close agreement between the dates derived by the different approaches strongly suggests their reliability.

One problem which has caused a great deal of confusion in early Pleistocene studies is the definition of the term 'Villafranchian'. This was first introduced during the last century for deposits laid down in the Piedmont area of Italy during the late Pliocene as the sea regressed, so that the boundary between the lacustrine Villafranchian deposits and the littoral sands underlying them is not of the same age in all places. However, the type Villafranchian does contain a characteristic fauna. Unfortunately, the geologist who introduced the Villafranchian confused this fauna with other faunas of later date, a practice continued by workers in other parts of Europe until a number of distinct mammalian faunal assemblages were all regarded as 'Villafranchian', and of early Pleistocene age. In fact, only 'middle' and 'late' Villafranchian faunas date from the early Pleistocene. The Villafranchian persisted until the beginning of the major continental glaciations, about 0·8 million years ago.

Early work in the Alps suggested that Europe had been subject to the advances and retreats of four ice-sheets, and though later studies revealed evidence of a fifth, earlier, Alpine glaciation, there has been a tendency among geologists to think in terms of series of four. Thus evidence of four glaciations was found in North America, and of four *pluvials*, or times of increased rainfall, in Africa. It has been claimed that the pluvials corresponded in time with the European glaciations, but though it now appears that the last pluvial and last glaciation were synchronous, the others were probably not.

It is misleading, moreover, to think of the major glaciations as unitary, discrete events, the ice advancing during long periods of colder conditions and retreating in warmer ones. Rather, the picture is a complex one of numerous fluctuations in extent of the ice-sheets. Also, glaciations were profoundly affected by local conditions, and a sequence worked out for one area cannot reliably be applied elsewhere.

However, the Alpine glacial sequence does provide a useful framework within which to place middle and late Pleistocene European hominids, especially since absolute dates from this time in this area are so scarce. This sequence is shown in Fig. 10.1. The major glaciations are separated by *interglacial* periods, and consist of colder periods, *stadials,* separated by warmer *interstadials.* Since one of the major geological results of glaciation consists of the erosion of the land surface by movements of ice, each glaciation tends to obliterate the geological evidence of preceding ones. Hence we know a great deal about the last glacial, the Würm, but very little about its predecessors. Most dates given for the major glaciations are based on palaeotemperature sequences deduced from deep-sea cores, but in the present confusion in European Pleistocene studies it is almost impossible to find two authorities who will agree on such dates. Thus about all we can do at the moment is to state that the first, Donau, glaciation began about 0·8 million years ago, and that the Würm began about 70 000 years ago. The duration of the Riss-Würm interglacial was probably around 30 000 years so we can date the end of the Riss at about 100 000 years ago.

Because of the great uncertainty surrounding the dating of the Pleistocene sequence, it seems best to refer to the 'early Pleistocene' corresponding to the middle and late Villafranchian, lasting from about 1·8 to 0·8 million years ago; the 'middle Pleistocene' starting at around 0·8 million and lasting until about 250 000 years ago; and the 'late Pleistocene', from 250 000 years ago to the present. There

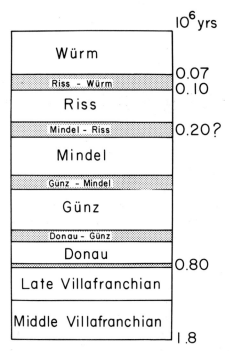

FIG. 10.1.   The Alpine glacial sequence of Europe.

have been some attempts to place the last 12 000 or so years in a separate epoch, the Holocene, or Recent, but such a division is quite unwarranted.

11 Late Pliocene and Early Pleistocene Hominids

South Africa

In 1925 most authorities accepted the Piltdown 'fossil' (later shown to be a fraud consisting of the braincase of a modern man combined with a suitably modified ape jaw) as representing the early Pleistocene forerunners of man. Piltdown man appealed to scholars of the time largely because of its large brain, then thought to be the most fundamental characteristic of man. Thus when, early in that year, Raymond Dart announced the discovery of an early Pleistocene hominid with a small brain, the general reaction of the anthropological world was one of disbelief. Dart's specimen, which he named *Australopithecus africanus* (African southern ape) consisted of the face, jaws, teeth and a brain cast of a juvenile at approximately the stage of development of a modern six-year-old human child. It had been found at the end of 1924 by workers in a lime quarry at Taung, in what is now Botswana, South Africa. In his initial description, Dart placed *Australopithecus* in its own family, Homo-simiadae, midway between the apes and man, but he did note that the fossil bore a far closer resemblance to man than to the apes. His judgement was strikingly vindicated when, in 1929, the lower jaw was finally freed from the mineral matrix which held it to the skull, and an unquestionably hominid dentition was revealed. The teeth of the juvenile are very large, but morphologically similar to those of later hominids.

In 1936, Dr Robert Broom, working at a site known as Sterkfontein, in the Transvaal, discovered many more individuals of *A. africanus*. Initially, these were placed in a new genus and species, *Plesianthropus transvaalensis*. Two years later, at Kromdraai, a site very close to Sterkfontein, Broom found the first evidence of a rather different kind of early hominid, which he named *Paranthropus robustus*. This form is rather more robust than is the gracile (delicate) *A. africanus*, but nowadays most authorities agree in placing *Paranthropus* as a separate species of *Australopithecus*, *A. robustus*. This is not to say, however, that the precise relationship of these two forms is anywhere near settled. Further finds of *A. africanus* were made by Dart in 1948 at Makapansgat, also in the Transvaal but about 150 miles north-east of Sterkfontein and Kromdraai. Under the mistaken impression that these hominids used fire, Dart initially named them *A. prometheus*. Finally, in 1949, Broom discovered the very prolific site of Swartkrans, close to Sterkfontein, which has yielded around 200 specimens of *A. robustus*. These were first described as *Paranthropus crassidens*, and they show a remarkable amount of variability in the size and shape of their jaws and teeth.

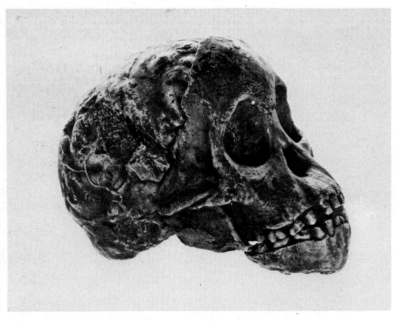

FIG. 11.1. Type specimen of *Australopithecus africanus*; infant skull from Taung. (Three-fifths natural size.)

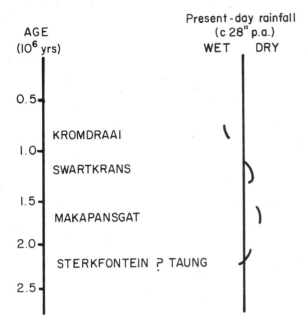

AGE
(10⁶ yrs)

Present-day rainfall
(c 28" p.a.)
WET   DRY

0.5

KROMDRAAI

1.0

SWARTKRANS

1.5

MAKAPANSGAT

2.0

STERKFONTEIN ? TAUNG

2.5

FIG. 11.2. Tentative chronology of the South African sites which have yielded *Australopithecus*. On the right are Brain's rainfall estimates for the sites.

Unfortunately, the South African sites are difficult to date, consisting as they do of accumulations of mineral and bone materials in caves formed in ancient rocks. Only Sterkfontein can be pinned down within any reasonable range, its fauna suggesting an age of about 2·0–2·5 million years. The other sites present a more difficult problem, but a tentative arrangement would place them as shown in Fig. 11.2. Dr C. K. Brain has inferred the climates prevailing at the times of deposition of the sites by an analysis of the sand grains composing the cave breccias. He believes that the

Sterkfontein cave was filled during a time somewhat drier than the present, and that Makapansgat represents an even drier environment. The rainfall at Swartkrans was about the same as it is in the same region today, while Kromdraai was considerably wetter. What sorts of environment do these rainfall estimates represent? In fact, a relatively small ecological variation: from savanna grassland to open woodland. It has been suggested that *A. robustus*, coming from the later, wetter, sites, was more herbivorous than *A. africanus*, but such a hypothesis is probably unjustified; the ecological range of baboons, for instance, encompasses a much wider variety of environments, and it seems unlikely that early hominids were any less ecologically variable.

The best-preserved skull of *A. africanus* is Sterkfontein 5, thought to have been female. The cranial vault of this specimen is high and rounded; its deep but delicate face is tucked in beneath the front of the braincase, and is surmounted by small brow-ridges. The face is rather prognathous (projecting), however, primarily to accommodate the large teeth. The cranial capacity of Sterkfontein 5 is about 480 cm³, very close to the mean volume derived from the three other Sterkfontein skulls sufficiently well preserved to allow this to be measured. This is almost the same as the average cranial capacity of the gorilla, and statistical calculation of the expected range of variation in the brain size of Sterkfontein *A. africanus* also closely coincides with the gorilla's, 99% of individuals of the Sterkfontein population being expected to have cranial capacities of between 250 and 720 cm³. As we have already seen, the small brain capacity of *A. africanus* was at first claimed to demonstrate its ape affinities, but besides the obvious fallacy it should be remembered that *A. africanus* was a great deal smaller than the gorilla, and that the organisation of its brain, as far as can be told, differed from that of apes—a far more critical feature than that of size.

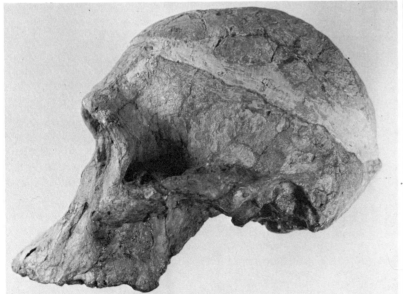
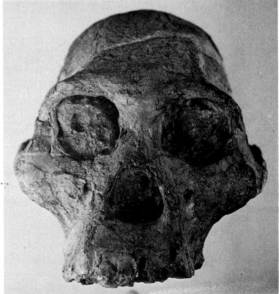

FIG. 11.3. Skull of *Australopithecus africanus:* Sterkfontein 5. (Three-fifths natural size.)
Courtesy of Transvaal Museum

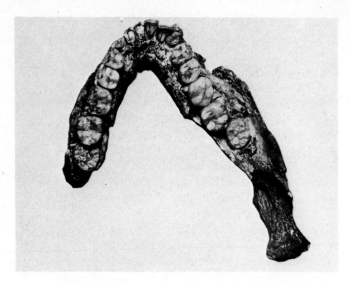

FIG. 11.4a. Mandible of *Australopithecus africanus* from Sterkfontein (Sts 52b).

Courtesy of Dr A. W. Crompton

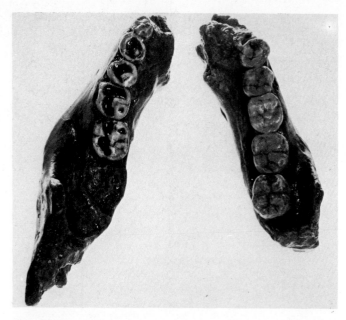

FIG. 11.4b. *Left:* mandible of *Australopithecus africanus* from Makapansgat. *Right:* type specimen of *'Paranthropus robustus'* from Kromdraai. This illustration may help to reveal why there is a certain degree of confusion in the study of early South African hominids. (Two-thirds natural size.)

In respect of its morphology and orientation, the dentition of *A. africanus* is very man-like, and its arrangement is close to the modern parabolic arcade although, as was also probably the case with *Ramapithecus*, the tooth-rows were longer, and rather squarer than those of modern man. The canines and incisors are small and vertically emplaced, while the lower premolars are homomorphic. In shape, the molar teeth are reminiscent of those of modern man, although, like the premolars, they are somewhat broader than in most modern populations. In size, and especially compared to body size, however, the cheek teeth of *A. africanus* are distinctly larger than those of later hominids, and even those of some pongids, and provide a considerable surface-area for grinding. This strong grinding adaptation is further emphasised by the fact that the hard enamel coating of these teeth is much thicker than that on modern hominid molars.

Certain of the features of the skull of *A. africanus* correlate with an upright posture. The occipital region of the skull has a rounder contour, and the nuchal area is much more horizontal than it is in apes, as is the plane of the foramen magnum. The occipital condyles, where the skull articulates with the vertebral column, are situated relatively far forward. All these adaptations are concerned with perfecting the balance of the skull atop the spine. Much more direct evidence of erect posture is of course provided by the postcranium, the lower half of which is quite well known from Swartkrans, the most valuable specimens being most of a vertebral column, a fairly complete pelvis and the proximal end of a femur. All of these belonged to a single individual who was lightly built, weighing probably no more than 70 lb, and who was not above 4 ft 6 in. in height.

The total morphological pattern of the spine, pelvis and femur of *A. africanus* is hominid: wedging of the lumbar vertebrae indicates the presence of lumbar curvature, as does the angling of the sacrum with respect to the innominate bones; the ilia are broad and short; the femur possesses a carrying angle. Of course, there are differences in detail from *Homo sapiens*: for instance, the ischium is slightly elongated, indicating that the hamstring muscles, extensors of the thigh and flexors of the lower leg, were of greater importance in the locomotion of *Australopithecus* than they are in modern man. The ilia curve round behind the acetabula, rather than in front of them as in modern man, suggesting that the gluteus medius and minimus muscles functioned rather differently, and the acetabula themselves face directly sideways instead of slightly forwards. But though differences do exist, it is clear that *A. africanus* was morphologically adapted to habitual erect bipedalism almost as well as we are.

As far as the shoulder girdle and upper limb of *A. africanus* are concerned, the picture appears to be somewhat different, based though it is on extremely limited fossil evidence. In his description of a fragmentary scapula and the proximal end of a humerus from Sterkfontein, Broom concluded that they were man-like, but that they did resemble those of the orang in some features. More recently, a statistical analysis of the scapula by Dr Charles Oxnard has confirmed the resemblance to the orang, and studies of the shoulder regions of a wide variety of living primates by the same author have suggested that the minimum evolutionary pathway from an arboreal shoulder girdle to that of man would have been from one similar to that of the orang. The Sterkfontein humerus is as big as that of a modern man, and belonging to a much smaller animal implies that the arms of *A. africanus* were relatively more robust than those of *Homo sapiens*, and that the two species may have differed in their bodily proportions. It is interesting that the

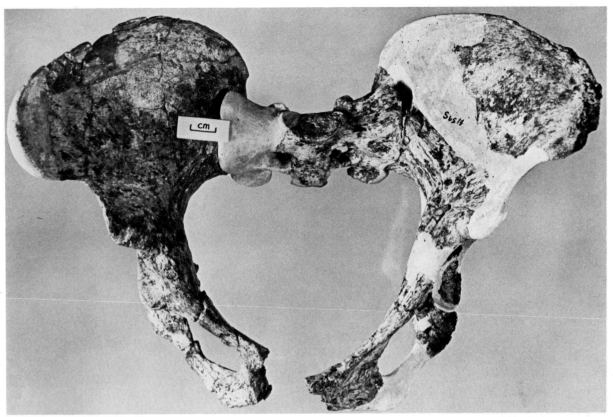

FIG. 11.5a. Postcranial remains of Sts 14, an adult female *Australopithecus africanus* from Sterkfontein. *Top:* front view of partially reconstructed pelvis (two-thirds natural size); *lower left:* lumbar portion of vertebral column articulated with left innominate bone; *lower right:* femoral shaft with partially reconstructed head (two-thirds natural size).

All courtesy of Dr J. T. Robinson

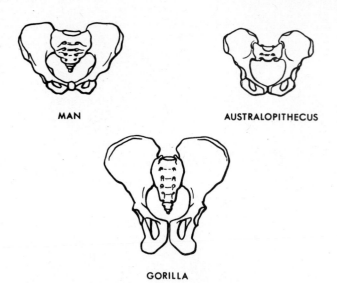

MAN   AUSTRALOPITHECUS

GORILLA

FIG. 11.5b. Comparison of the reconstructed pelvis of *Australopithecus africanus* from Sterkfontein with the pelves of man and the gorilla.

heads of the humeri of living primates which habitually suspend themselves from their arms are large; the possession in *A. africanus* of a large humeral head might be held to support Oxnard's views. However, perhaps it is better to suspend judgement on this until more and better material is known.

A recent demographic study of *Australopithecus* by Dr A. E. Mann has produced some interesting conclusions. The timing and sequence of dental eruption in *Australopithecus* seems to have been like that of modern man, and in distinct contrast to that of the apes. This implies that the entire developmental sequence of *Australopithecus* was like that of *Homo sapiens*; the most important inference from this is that the length of time of dependence of juveniles upon their mothers, a time during which a multitude of indispensable social skills are acquired, was long. Further, Mann has shown that the average age at death among *Australopithecus* was around 18 years, and that few individuals survived the age of 40. Thus we may infer that if *Australopithecus* started breeding at the age of 13, an age comparable to that among many primitive societies today, it would have been unusual for an individual to have both parents, or even one, alive when he reached maturity. The obvious corollary of this is that *Australopithecus* must have possessed a social organisation of sufficient complexity to cope with the socialisation of orphaned juveniles.

What else can we say of the social organisation of South African *Australopithecus*? Stone tools have been found at a number of the South African sites, but since no living-floors have yet been found, we can say little about the complexity of his tool-kit. *Australopithecus*, or at least the later members of the lineage, probably made his living as a hunter-gatherer, much as do some primitive tribes today. It is perhaps appropriate to point out that the role of hunting in a hunting and gathering society has generally been over-

estimated; hunting accounts for less than 25% of the food intake of such peoples. Nevertheless, it is probable that there was some division of labour among *Australopithecus*, the females gathering vegetable food, and the males hunting small game in a co-operative manner. Again by analogy with hunting-gathering societies, *Australopithecus* would have been wide-ranging, living in small, mobile groups. In many reconstructions *Australopithecus* is depicted as having been covered in hair, but it should be noted that hominid bipedalism, though efficient, is not fast, and that the typical hominid hunting pattern depends on the continuous expenditure of energy, and therefore production of heat, over long periods, perhaps days in extent. A hairy coat prevents the efficient dissipation of heat, so that it is reasonable to assume that if *Australopithecus* indulged in diurnal hunting activities, as he probably did, he was not the hairy beast of the cartoons.

We have discussed the morphology of gracile *Australopithecus*. What of the robust form from South Africa? Probably the behavioural speculations above apply to *A. robustus* as much as to *A. africanus*, but there are dental and osteological differences which require discussion. Unfortunately, the robust material is not preserved so well as that of *A. africanus*, and brain volumes cannot be so reliably computed. However, the cranial capacity of *A. robustus* is unlikely to have differed greatly from that of *A. africanus*.

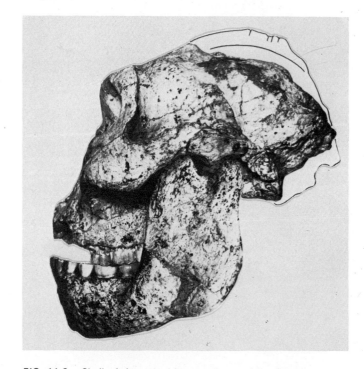

FIG. 11.6. Skull of *Australopithecus robustus* from Swartkrans; the cranium and mandible do not belong to the same individual.
Courtesy of Dr E. L. Simons

Most of the cranial distinctions between the gracile and robust forms can be explained in terms of their dental differences. The face of *A. robustus* is relatively flat, due largely to the fact that its anterior teeth are very small. In direct contrast are the cheek teeth, which are massive, being on average 25% greater in occlusal area than those of *A. africanus*. The premolars have become incorporated into the molar field; the anterior one is large, and the posterior completely molariform. These big teeth are implanted in massive jaws which require a powerful masticatory musculature; the face is therefore strongly built to resist and resolve great chewing stresses. Sagittal cresting in some individuals results from the combination of large temporal muscles with a small braincase. Among other differences from *A. africanus* also probably explicable in terms of the

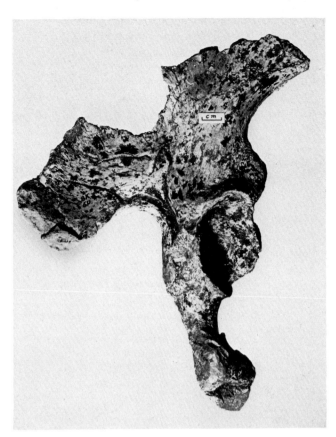

FIG. 11.8. Innominate bone of *Australopithecus robustus* from Swartkrans. (One-half natural size.)   *Courtesy of Dr J. T. Robinson*

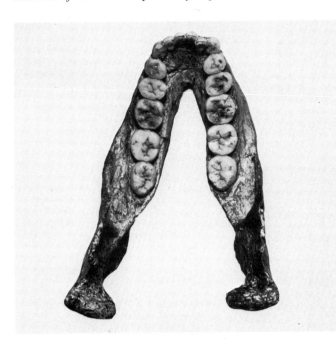

FIG. 11.7.   Mandible of *Australopithecus robustus* from Swartkrans (Sk 23). The specimen has suffered from lateral compression during fossilisation.

Courtesy of Dr A. W. Crompton

dental mechanism is the situation of the face in front of, rather than below, the anterior part of the braincase. This results in a rather flatter frontal area and the possession of somewhat more pronounced brow-ridges.

The postcranial anatomy of *A. robustus* is very little known, our knowledge of it being restricted to a couple of metacarpals, a distorted and incomplete innominate bone and the upper ends of two femora from Swartkrans, and a talus from Kromdraai. The Swartkrans pelvis, fragmentary and distorted though it is, clearly differs from that of *A. africanus*, but the meaning of these differences remains obscure. The pelvis is that of an upright biped a good deal larger than *A. africanus*, and much heavier; its greater body size may go some way towards explaining the dental and cranial differences between the gracile and robust forms.

The *A. robustus* sites, Swartkrans and Kromdraai, are rather younger than those classically yielding *A. africanus*. However, at one of these sites, Swartkrans, there does appear to be evidence of co-existence between the two species. In 1949, Broom and Dr J. T. Robinson described a mandible from this site, naming it *Telanthropus capensis*.

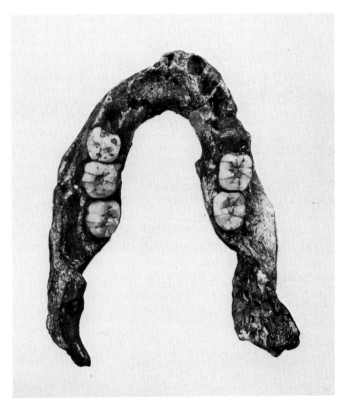

FIG. 11.9.   'Telanthropus' from Swartkrans.
Courtesy of Dr A. W. Crompton

A few other fragments of this form have since been found. The mandible is much more delicately built than is typical of *A. robustus*, is smaller in size, and has smaller teeth. It is closest in morphology to *A. africanus*, but bears some unique features, for instance a total lack of buttressing at the front of the jaw, either in the form of a simian shelf or of a chin. Some authorities have claimed that *Telanthropus* should be assigned to *Homo erectus*, man's middle Pleistocene fore-

African *Australopithecus* merely represented local variants of a widespread, and presumably polytypic, genus. Nevertheless, it was not until 1959 that evidence began to accumulate of the presence of *Australopithecus* elsewhere.

Most of this evidence comes from Olduvai Gorge in northern Tanzania, where a river has cut through a sequence of deposits, partly of lacustrine origin, covering most of the Pleistocene. Periodic volcanic activity has permitted us to

OLDUVAI GORGE

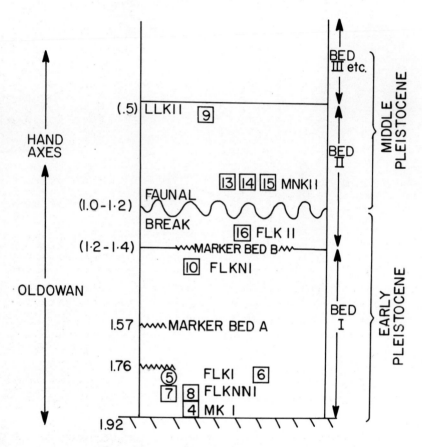

FIG. 11.10. Chart showing the sequence of fossil hominids at Olduvai Gorge. Dates shown in brackets are tentative. The hominids are identified by their numbers. All hominids whose numbers are enclosed in squares are considered to belong in the same lineage; hominid 5, represented by a circle, is the only member of its lineage so far found at Olduvai.

runner, and the next arbitrary species in our continuum, to which the mandible does show some similarities. This suggestion has recently been reinforced by the reconstruction by R. J. Clarke of a partial face and braincase, probably referable to the same species, which appears to have some affinity to *H. erectus*.

East Africa

Despite the fact that there was a tendency for a long time to regard *Australopithecus* as a purely South African phenomenon, it was quite clear that both gracile and robust South

make a number of absolute age determinations on these deposits. The sedimentary series has been divided into five beds, the bottom of the earliest, Bed I, having a K-Ar date of about 1·9 million years, and therefore coinciding closely with the base of the Pleistocene. Together with the lower part of Bed II, Bed I covers the lower Pleistocene; upper Bed II and Beds III and IV are of middle Pleistocene age; Bed V is of recent origin.

For some years Beds I and lower II at Olduvai had been known to contain crude stone tools known as *Oldowan*, and the excavation of complete living-floors by Dr and Mrs L. S. B. Leakey has revealed a remarkable diversity of these

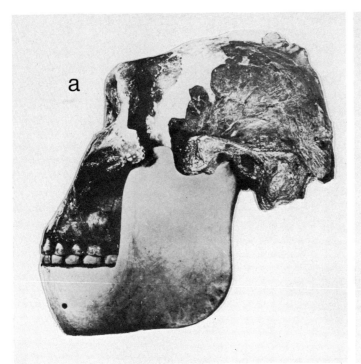
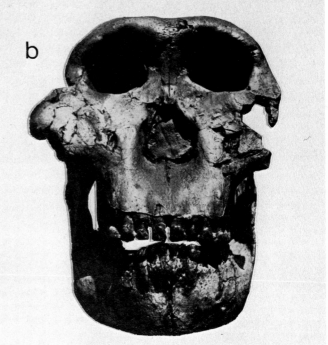

FIG. 11.11.   Cranium of *Australopithecus boisei*, Olduvai hominid 5. (*a*) lateral view, articulated with hypothetical mandible; (*b*) frontal view, articulated with the Peninj mandible.
(*a*) *Courtesy of Dr E. L. Simons;* (*b*) *courtesy of Dr L. S. B. Leakey,* © *National Geographic Society*

implements, although the manufacture of standard tool types was little developed; perhaps one fact may help to explain the other. But there was no direct evidence of the existence at Olduvai of hominids during the early Pleistocene until, in 1959, Mrs Leakey discovered the almost complete skull of a robust hominid (Olduvai hominid 5) near the bottom of Bed I, and therefore about 1·5 million years old. With a cranial capacity of about 530 cm³, this skull in many ways resembles that of South African robust *Australopithecus*, but it is even more heavily built. Its cheek teeth are huge, being considerably larger in surface area than those of

A. robustus, and over twice the size of those of *A. africanus*. Both premolars are large and molarised, but the canines and incisors are extremely small, and vertically implanted. In correlation with the massive molar-premolar series, the face and braincase of this hominid are very strongly built, far more so than in *A. robustus*. The face of hominid 5 is also a great deal deeper than that of the South African form. All these features demonstrate an extremely powerful chewing mechanism, more developed than in any *A. robustus*.

Leakey first described this hominid as the type of a new genus and species, *Zinjanthropus boisei*. A more recent study

FIG. 11.12.
Palate and upper dentition of Olduvai hominid 5 (*left*) compared with upper dentition of Olduvai hominid 16.
(One-half natural size.)
    Courtesy of L. S. B. Leakey,
    © *National Geographic Society*

by Dr Phillip Tobias concludes that all the African material belongs in genus *Australopithecus*, but that Olduvai hominid 5, which bears resemblances in detail to both *A. robustus*, and to *A. africanus*, should be classified in its own species, *A. boisei*. But Tobias does believe that *A. boisei* is ancestral to *A. robustus*. Pertinent to discussion of this view is a mandible found at Peninj, near Lake Natron, also in Tanzania. Originally thought to have been of early middle Pleistocene age, the specimen is now considered to be in excess of 1·4 million years old. This lower jaw is large and robustly built, though not as massive as the *A. boisei* mandible from Ethiopia, which we will discuss shortly, and is reminiscent of the Olduvai and Ethiopian robust hominids in the relative development of its anterior and posterior teeth. However, it does differ from *A. boisei* in a number of

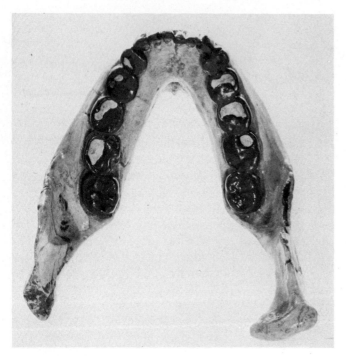

FIG. 11.13.   The Peninj mandible. (Three-fifths natural size.)

respects, for instance in the relative breadth of its teeth, which are considerably narrower, and in the height of its ascending ramus, which is much lower than that of Olduvai hominid 5 would have been, as inferred from the depth of its face. Dr Tobias has referred this mandible to *A. robustus* on his assumption that the robust East and South African material represents a single lineage characterised by evolutionary reduction in the posterior dentition and in facial depth, but although the Peninj jaw is now known to occupy a time probably intermediate between the South and East African robust forms, it is difficult to visualise the fossil as being morphologically intermediate between the two groups, although the possibility cannot be ruled out entirely. Certainly the Peninj mandible is generally more similar to the *A. boisei* material than to any of the South African material, while in many ways the gracile and robust

South African hominids resemble each other more closely than either does *A. boisei*, but it may be that here we have evidence for yet another lineage of hominids in the early Pleistocene of East Africa, particularly since the Peninj specimen does show some similarities to the problematical partial mandible from Locality 17 at Omo, in Ethiopia, which we will discuss later.

In 1959 *A. boisei* was immediately acclaimed as the manufacturer of the Oldowan tools of Bed I, Olduvai. But two years later its claims were supplanted by those of a newer candidate, in the form of the hominid represented by a juvenile mandible which came to light near the base of Bed I, at a level slightly below that at which Olduvai hominid 5 had been found. In point of fact, there is little to be gained by quibbling over which hominid made the Oldowan tools; presumably either would have been capable of this, and two early hominids, manufacturing implements of so unstandardised a kind, might easily have been making almost indistinguishable tools.

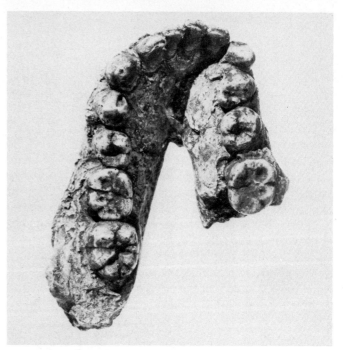

FIG. 11.14.   Type mandible of 'Homo habilis' from Bed I, Olduvai. (Natural size.)

The new mandible, Olduvai hominid 7, differs markedly from robust *Australopithecus*, resembling far more closely *A. africanus* from Sterkfontein, although there are distinctions even between these; for instance, the South African form has rather broader premolars, and its teeth are more closely packed. Olduvai hominid 7 was described in 1964 as the type of a new species, *Homo habilis*. At the same time, certain other fossils from the bottom of Bed I were referred to the same taxon. These included a pair of skull bones—the parietals—and some hand bones, probably belonging to the same individual as hominid 7; a partial adult hand; and an almost complete foot (Olduvai hominid 8). At the site where the *A. boisei* skull was found were also recovered a delicate

FIG. 11.15.   Parietal bones of 'Homo habilis' from Bed I, Olduvai Gorge; white areas reconstructed. (Two-thirds natural size.)

tibia and fibula, most appropriately ascribed to *Homo habilis*. The foot is unquestionably adapted to erect bipedalism, and although small, closely resembles that of modern man. Longitudinal and transverse arches were present, formed and supported as in ourselves. The big toe was robust, and set in line with the long axis of the foot. Unfortunately, all the toe bones are missing from this specimen, but an isolated terminal phalanx of the big toe

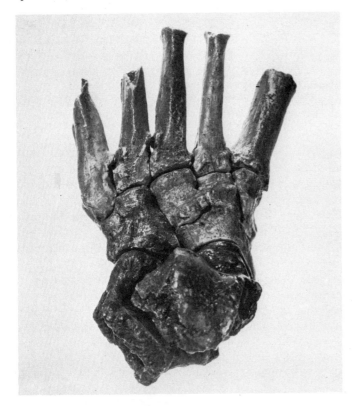

FIG. 11.16.   Foot of 'Homo habilis' from Bed I, Olduvai Gorge (hominid 8). (Natural size.)

(Olduvai hominid 10) was found at a site rather higher in the section, in association with Oldowan tools. As it happens, this is an extremely diagnostic bone, and its possessor was without doubt an habitual erect biped. The tibia and fibula, too, bear evidence of the bipedal habit; the individual of which they formed part was lightly built, and can have been little more than 4 ft tall. The hand of *Homo habilis* was not vastly different from our own, and one of the bones, the capitate, though badly preserved, is similar to a specimen from Sterkfontein.

The parietal bones of *Homo habilis* have been used in attempts to assess the brain volume of the complete skull, although with only these bones available, such assessment is a very hazardous business. The latest of a series of estimates is 657 cm³, a little outside the observed upper limit in *A. africanus*, though within the calculated limits for this group.

How should we classify *Homo habilis*? Clearly, the most suitable material for comparison is the penecontemporaneous *Australopithecus*, rather than the later *Homo*, and there seems little reason to exclude the Olduvai material from the former genus. As we have remarked earlier, classification at this level is largely a matter of taste, and the differences between *Homo habilis* and *A. africanus* might equally well be regarded as those between closely related species, or as merely representing variations between geographically and temporally separated populations of a single polytypic species. Clearly, there is no justification for generic separation, and I incline to the view that, until new material shows otherwise, *Homo habilis* is best regarded as a variant of the polytypic species *A. africanus*.

Before we discuss material from higher in the section at Olduvai, let us briefly consider some very recently acquired evidence for even earlier occupation of East Africa by *Australopithecus*. In 1965, Dr Bryan Patterson discovered the lower end of a humerus at Kanapoi, in north-west Kenya. This fragment, which multivariate analysis has shown to be hominid, is believed by its describers to have belonged to gracile *Australopithecus*, although it is of large size. A lava flow overlying the deposits in which the humerus was found has been K-Ar dated at 2·5 million years, and the specimen was originally thought to have been of around that age. Now, however, it is known that the lava flow was intruded into the sediments at some time after they had been laid down, and the Kanapoi fauna suggests an age of about 4 million years, since it is similar to that found at the bottom of the series of sediments in the Omo basin of south-west Ethiopia, which covers the period between 4 and 2 million years ago.

The Omo beds have been worked since 1967 by an international team with contingents from France, Kenya and the U.S.A. Isolated hominid teeth, most of them similar to those of gracile *Australopithecus*, have been found throughout the section. Recently, the American group, under the direction of Dr F. C. Howell, recovered two lower jaws from deposits near the top of the section, and therefore about 2 million years old. One of these, lacking only the ascending rami, is

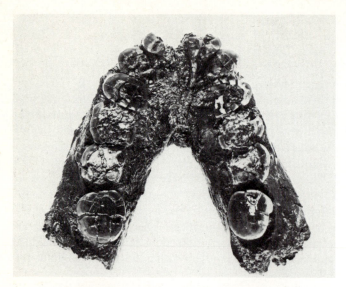

FIG. 11.17.   The *Australopithecus boisei* mandible from Locality 7, Omo. (Two-thirds natural size.)

Courtesy of Dr F. C. Howell

the most massively built hominid mandible yet found, and almost certainly belongs in the lineage of *A. boisei*, with which it is approximately contemporaneous. The anterior teeth are small, but the molars and premolars are very large, and just as one would expect the lower teeth of Olduvai hominid 5 to look. The other jaw is more fragmentary, consisting of a right horizontal ramus containing only the canine and the last premolar; the anterior premolar has made contact facets on both these teeth, however, and the distance between them indicates an extreme elongation of this tooth, quite unlike that in any other *Australopithecus* yet known. Generally, this specimen is most reminiscent of the Peninj jaw. In 1967 the French contingent recovered a mandible, lacking all tooth crowns, from a site below a tuff K-Ar dated at around 2·5 million years. This specimen again differs in some respects from previously known material, but is most plausibly interpreted as belonging to an early member of the *A. boisei* lineage.

The earliest evidence of *Australopithecus* consists of a

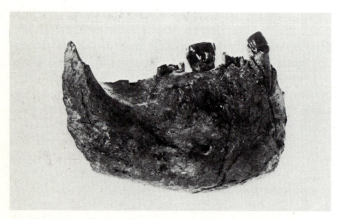

FIG. 11.18.   Partial *Australopithecus* mandible from Locality 74, Omo. (Two-thirds natural size.)

Courtesy of Dr F. C. Howell

partial lower jaw found by an expedition under the direction of Dr Patterson at Lothagam Hill, in Kenya. This site lies stratigraphically below Kanapoi, and is probably in the region of 5 million years old. The jaw, containing a single tooth, is large, and might best be regarded as representing an early gracile *Australopithecus*; the remaining (first) molar is squarish, and most reminiscent of those from Makapansgat, while their roots show the second and third molars to have been very large.

Meanwhile, back at Olduvai Gorge, some fossils from Bed II were assigned to *Homo habilis* when the taxon was named. These include a mandible, maxillae and some broken skull bones (occipital, parietal and temporals), all of which may belong to the same individual (Olduvai hominid 13);

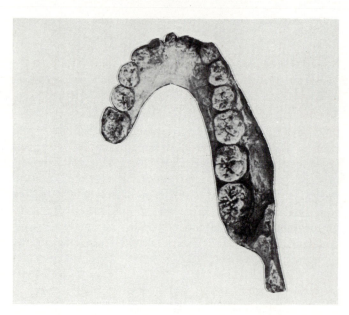

FIG. 11.19.   Mandible of Olduvai hominid 13, from Bed II, Olduvai Gorge. (Three-quarters natural size.)

Courtesy of Dr L. S. B. Leakey; © *National Geographic Society*

some isolated teeth (hominid 15); and some juvenile skull fragments (hominid 14). All these were found at a site immediately above a faunal break in the Olduvai sequence which is probably a little over a million years old and represents the transition at Olduvai from early to middle Pleistocene. Below this faunal break, but above the upper limit of Bed I and probably about 1·3 million years old, was found another skull, Olduvai hominid 16, which was provisionally referred to *Homo habilis*.

Since their description there has been a great deal of controversy over these finds. Starting with the earliest, hominid 16 has a cranial capacity of about 640 cm³, and a rounded and lightly built skull reminiscent of that of gracile *Australopithecus*. This is in contrast with *Homo erectus*, which has a long, relatively flattened, and thickly walled skull. The dentition of this form, however, is distinctly reduced compared to that of the Bed I material, and approximates much more closely to that of *Homo erectus*. In fact, hominid 16 provides us with a good intermediate between the gracile Bed I form and the long, heavily built skull of 'Chellean Man' (hominid 9) from the top of Bed II, which is clearly *Homo erectus*. We have already remarked upon the difficulty of classifying intermediate forms; how should hominid 16 be classified? But first let us look at the other Bed II material.

Hominids 13, 14 and 15 present us with the same problem. The hominid 14 skull again resembles that of gracile *Australopithecus*, while the teeth resemble those of *Homo erectus* from Java. We have, therefore, three choices open to us in classifying this material. If we wish to use a time-boundary, we could most conveniently place this at the early-middle Pleistocene division. Hominid 16 would therefore be assigned to gracile *Australopithecus*, and hominids 13, 14 and 15 to *Homo erectus*. Alternatively we could accept the evidence of the dentition as being most meaningful, and place all the Bed II material in *Homo erectus*. Or we could place most emphasis on the skull and regard the Bed II specimens as advanced gracile *Australopithecus*. Of these alternatives, my preference at present would be to regard the Bed II material as representing a stage in the lineage leading from Bed I '*Homo habilis*' to 'Chellean Man', which, though advanced, is not sufficiently evolved

to warrant generic separation from gracile *Australopithecus*.

Throughout Bed I and most of Bed II at Olduvai, artefacts are of the Oldowan 'pebble tool' type. Only towards the top of Bed II does a more sophisticated type of implement, the 'hand-axe' appear. The more evolved culture of which the hand-axe is typical is known as the *Acheulean*, and began about 0·5 or 0·75 million years ago. In North and South Africa the first appearance of the Acheulean is intrusive, presumably from the region of Central or East Africa; pebble-tools persisted much later in Europe than in Africa, and advanced industries appeared in Asia only late in the Pleistocene. Dr Glynn Isaac has suggested that the Acheulean culture originated with the discovery of a technique of striking flint flakes more than 10 cm long, and that this technique spread only slowly by cultural transmission. The changelessness of the Oldowan culture over a period of well over a million years coincides with a period of relatively static hominid brain size, and suggests that *Australopithecus*, relatively unintelligent compared to later hominids though it was, was relatively well adapted to prevailing conditions. Only, it appears, with the arrival of the middle Pleistocene did tool-making acquire a greater cultural component at the expense of the biological one.

12   Middle Pleistocene Hominids

The first discovery of *Homo erectus* was presented to an unreceptive world in 1891 by Eugene Dubois, who had recovered a fossil skullcap from alluvial sediments on the banks of the Solo River in Java. The middle Pleistocene deposits of Java fall into two faunal zones, the earlier Djetis and the later Trinil. There is a K-Ar date on a lava flow at the top of the Trinil of about 0·5 million years, while tektites at the Djetis-Trinil boundary have been dated at 710 000 years. The Djetis beds may extend back to as much as a million years ago.

Homo erectus material from the Djetis beds is less plentiful than that from the Trinil, and consists of parts of the face and skull of a single individual, two incomplete mandibles and the braincase of an infant perhaps two years old. The adult material possesses classical *Homo erectus* features: a long, low, wide, thickly walled skull with a flattened frontal area, large brow-ridges and a sharply angled occiput, combined with small teeth arranged exactly as in modern man.

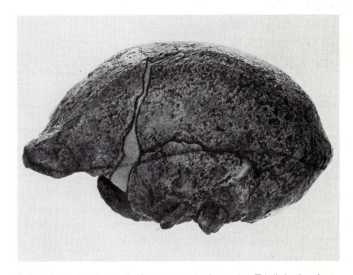

FIG. 12.1   Skullcap of *Homo erectus* from the Trinil beds, Java. (One-half natural size.)

The infant incipiently shows the same cranial features, although, unsurprisingly, its skull is thin-walled. In a comparative study, Drs Phillip Tobias and G. H. R. von Koenigswald were impressed by the dental resemblances between this material, Olduvai Bed II *'Homo habilis'*, and *'Telanthropus'* from Swartkrans; however, in the conformation of its skull, Olduvai hominid 13 differs considerably from the Djetis specimen. Its braincase is lighter, shorter and higher than the latter's, and its brain is a bit smaller (620 cm³ as opposed to around 750 cm³). The evolution of a polytypic lineage spread widely over the Old World would obviously have been a very complex affair, and since our knowledge of the hominid fossil record is so widely scattered in place and time, we cannot at present make any strong guesses as to the exact relationships between the African and Asian forms. However, it is likely that different populations were evolving mosaically and in slightly different ways during the middle Pleistocene, and the African material may on present evidence plausibly be regarded as representing a stage ancestral to that represented by Djetis *Homo erectus*, though not as being directly ancestral to the Java group.

Just how complicated the picture was is shown by the presence in the Djetis beds of a form initially described as *'Meganthropus palaeojavanicus'*, and represented by three

FIG. 12.2. Mandibular fragment of *'Meganthropus palaeojavanicus'* from the Djetis beds, Java. (Natural size.)

mandibular fragments. These specimens are massive, and bear large teeth. Tobias and von Koenigswald regard *Meganthropus* as representing one 'grade of hominisation' below the other Djetis material, a grade also occupied by (presumably Bed I) *Homo habilis*; other workers have assigned it to either robust *Australopithecus* or *Homo erectus*. The last of the three assignments may provisionally be regarded as the most satisfactory; the skull of *Meganthropus* is as yet unknown, and in comparable parts the distinctions between all these forms are subtle.

From deposits at Lantien in north-west China thought to

be of Djetis-equivalent age have come a calvaria (braincase) and a lower jaw. The calvaria is closely similar to that of Djetis *Homo erectus*, with a brain volume of about 780 cm³; it is long, low and thick, with an angulated occiput and large brow-ridges. The mandible is the earliest recorded to lack third molars, a trait which occurs with a frequency of up to 30% in some modern populations.

Dubois' original material is of Trinil age, and consists of a calotte (skullcap) and a femur. The femur is interesting in that it shows no substantial differences from that of modern man; on the upper third of the shaft there is a remarkable pathological outgrowth. The initial response to this fossil was to regard it as a modern femur which had somehow found its way into the middle Pleistocene deposits, but chemical (fluorine) tests subsequently showed it to be of the same age as all the other Trinil material, and further femora from the same zone confirm that its human conformation is no freak. Assuming the proportions of *Homo erectus* to have been approximately the same as those of modern man, the femur belonged to an individual of about 5 ft 7 in. in height. This indicated increase in body size may go some way toward explaining the morphological differences between the crania of late gracile *Australopithecus* and early *Homo erectus*, but it is certainly not the whole story. Brain size did not differ greatly between the two groups, but cranial shape is very distinctly different, the relative flattening of the skull in *Homo erectus* being due to its lengthening and broadening, and the consequent flattening of the frontal area leading to the development of large brow-ridges to resolve chewing stresses transmitted through the face, and perhaps to serve as the site of attachment of large temporal muscles. But why the skull of *Homo erectus* became thus elongated remains a puzzle. The Trinil calotte is similar to those from Lantien and the Djetis. The frontals terminate anteriorly in large brow-ridges, behind which there is a pronounced lateral 'post-orbital constriction', a concomitant of combining a large face with a small, backwardly set braincase. The contour of the cranial vault is not evenly rounded from side to side, but shows a distinct fore-and-aft keel in the midline. Its cranial capacity is rather greater than that of the earlier specimens, at around 900 cm³. Originally Dubois named his finds *Pithecanthropus erectus*, but it is now universally agreed that there is no evidence at all to justify excluding the form from *Homo*. Further mandibular and cranial remains are known from Java, but they do not differ materially from those described above.

Homo erectus is most abundantly known from the cave at Choukoutien, near Peking, some 14 crania and 11 mandibles having been recovered from this site. These fossils were first described in 1927 as *Sinanthropus pekinensis*; subsequently it was realised that they were not generically distinct from Java man, and they became *P. pekinensis*. Now they are generally held to be distinct only at the subspecific level, and are known as *Homo erectus pekinensis*. It has been customary to give each geographical variant of *Homo erectus* a separate subspecific designation; however, it is unusual to give the geographic races of modern man formal

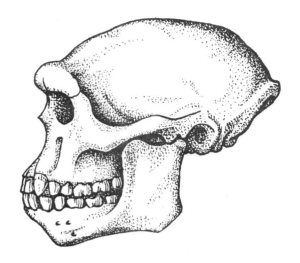
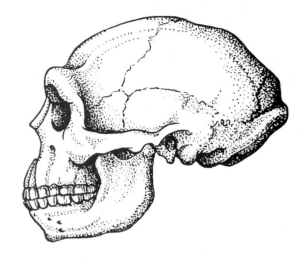

FIG. 12.3. Comparison of the reconstructed skulls of Trinil (*left*) and Lower Cave, Choukoutien, *Homo erectus*.

subspecific names, and it is doubtful whether the practice is of much utility when applied to *Homo erectus*.

The Peking skulls are probably a little younger than those from Java; the only means of dating them is by faunal correlation, which is difficult because there are no nearby faunas which can be absolutely dated. However, a date of around 450 000 years is unlikely to be too inaccurate.

Perhaps because of their later date, the cranial capacity of the Peking skulls is a bit larger than that of the Java crania, averaging sometimes over 1000 cm³, with a range of from 900 to 1200 cm³. Because of their larger brains, the skulls give the appearance of being rather better filled: the brain-cases are a little higher and more rounded, and midline keeling has disappeared. The brow-ridges are somewhat smaller because of the higher frontal region, and the occiput is more rounded, though it still retains a sharp angulation. Reconstructed, the face is large, and robustly built; it is prognathous, but not quite to the extent seen in von Koenigswald's reconstruction of the face of Java man. The mandibles possess a narrow, rounded dental arcade and lack a chin; instead, the symphysis recedes from the plane of the incisor teeth, being buttressed in compensation on the inside. The teeth are robust in comparison with those of modern man, but they are reduced compared to *Australopithecus*.

A few postcranial fragments are known from Choukoutien; these are morphologically identical to comparable parts of *Homo sapiens*, but are more strongly built.

There is evidence at Choukoutien that Peking man made crude 'chopper' tools from imported stone, and his use of fire is evident from the presence in the cave of ash and charcoal deposits; this is the first recorded case of the use of fire by man. From evidence at Choukoutien we may infer that *Homo erectus* was an accomplished hunter of big game; division of labour was probably practised, the males hunting and the females gathering vegetable food.

The same time-period has also yielded evidence of the presence of *Homo erectus* in both Africa and Europe. The earliest such evidence is the skull of 'Chellean man' (Olduvai hominid 9), discovered by Leakey in 1960 near the top of Olduvai Bed II, and probably about half a million years old.

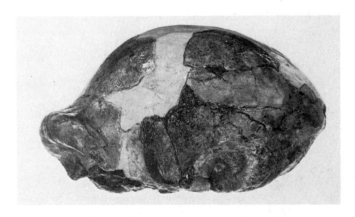

FIG. 12.4. Calvaria of 'Chellean Man' (hominid 9) from the top of Bed II, Olduvai Gorge. (Two-fifths natural size.)

This specimen consists of a thick-walled calvaria with a moderately low vault, angulated occiput and enormous brow-ridges. Its cranial capacity is in the region of 1000 cm³. Nearby were found hand-axes of a relatively advanced type. Of about the same age, or possibly a bit younger, are three mandibles and a parietal bone found at Ternifine, in Algeria. These are all robust and differ only slightly from comparable material at Choukoutien. Associated stone tools are not unlike those from upper Bed II at Olduvai.

A single lower jaw was found in 1907 during excavation of a sandpit at Mauer, near Heidelberg. Dating of the sand deposits is uncertain, but faunal associations suggest that a date of around 400 000 years may not be too inaccurate. The mandible is large and robust, with very broad ascending rami. Finally, a recent discovery of some teeth and an occipital bone has been made at Verteszöllös, in Hungary.

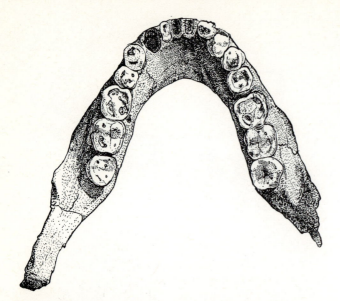

FIG. 12.5.   Mandible of *Homo erectus* from Ternifine, Algeria. (Four-fifths natural size.)

The site probably dates from late in the Mindel glacial, perhaps 400 000 years ago, and contains implements of the primitive 'chopper' type. The occiput is thick, though less sharply angled than is typical of *Homo erectus*, and has been described as belonging to an early *Homo sapiens* with an estimated brain volume of 1400 cm^3 (around the modern average). A more reasonable interpretation, however, would place the specimen in *Homo erectus*; the angulation of the occiput is rather greater than in most *Homo sapiens*, and a brain volume computed from a single occipital without knowledge of the shape of the rest of the skull can hardly be regarded as definitive. Probably, however modern the occiput, the rest of the skull was quite unlike that of *Homo*

sapiens; the isolated teeth from Verteszöllös are clearly more like those of *Homo erectus*.

Thus, summing up, *Homo erectus* was widely distributed over the Old World during the middle Pleistocene, known material covering the period from around a million until about 350 000 years ago. The species shows several advances over *Australopithecus*; notably, body size increased to modern proportions and the erect striding gait was fully achieved. The face, mandible and braincase were still robust, indicating a powerful masticatory action, although the teeth were reduced. In early *Homo erectus* the brain was not much larger (naturally enough) than in late *Australopithecus*, but for reasons which are still obscure it was accommodated in a skull of a different shape. During the middle Pleistocene, unlike the early part of the epoch, there was a steady increase in hominid brain size, which very probably correlated with increasing social and cultural complexity. In the very earliest stages, brain volume increase may merely have accompanied the enlargement in body size, but subsequently this cannot have been the case. Fossil remains of *Homo erectus* are accompanied by evidence of some technological advance, although in Europe and Asia tools, by and large, remained primitive; and the hunting of large animals in a manner which must have been co-operative certainly demanded a social structure of some complexity. The evidence of fire from Choukoutien and Verteszöllös recalls Dr E. R. Leach's suggestion that the use of fire (apart from its obvious material advantages) serves another, symbolic function: to proclaim the 'otherness' of the animal world. It is further likely that some form of language was employed by *Homo erectus*; without the use of language, the capacity to evoke and utilise abstract symbols in communication, even a rudimentary culture of the uniquely human sort is impossible.

13   Late Pleistocene Hominids

The late Pleistocene is the time from which archaic hominids are most abundantly known; unfortunately, however, this abundance has not helped to clarify the evolutionary picture, which was obviously complex, but is at present incompletely understood. A further complicating factor has been the geographical distribution of late Pleistocene hominids, a disproportionate number of which are known from Europe. This distribution reflects little more than the abundance of European fossil collectors, but none the less Europe has loomed unwarrantedly large in the minds of palaeoanthropologists when in fact, in both the geographical and evolutionary senses, it is peripheral. Men of the late Pleistocene prior to the emergence of the modern type can be generally characterised as possessing long, low, thickly walled skulls with large faces, brow-ridges and projecting occiputs, but with large cranial capacities. In all other morphological respects, they were as modern as we are, and it is now generally agreed that it is better to emphasise their similarities to modern man, rather than the differences, and to accord them separate subspecific status within our own species, as *Homo sapiens neanderthalensis*. Archetypal of these men are the Neandertals of the last interglacial and glacial of Europe. Claims have been made, however, that some fossils contemporaneous with, or earlier than, these are in fact representative of a pre-existing 'early sapiens' type. Since most of these specimens are of European provenance, it seems appropriate that we should begin our discussion of late Pleistocene hominids with Europe.

The earliest fossils for which such 'early sapiens' claims have been made come from the Mindel-Riss interglacial, perhaps 200 000 years ago. In 1935 a braincase, with most of the face attached, was found at Steinheim, West Germany. In most features it is of the type described above: the calvaria is long, narrow and not very high, with volume of about 1150 cm³; the face is large, if not very prognathous, and the brow-ridges are pronounced. But the occipital is rounded, and on the basis largely of this feature, the skull has been held to be of modern aspect when compared to the later 'classic' Neandertals. But in other respects the Steinheim skull makes a good intermediate between *Homo erectus* and later European hominids, and there is no good reason for claiming it to be modern on the basis of a single feature. In the same year, and from gravel deposits of similar age at Swanscombe, Kent, an occipital bone was recovered. In 1936 and in 1955, thick parietal bones belonging to the same

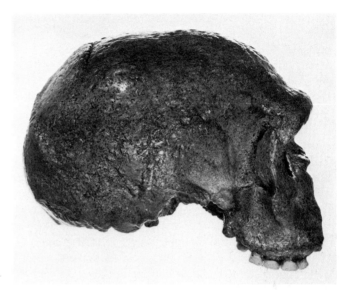

FIG. 13.1.   The Steinheim skull. (45% natural size.)

individual were found, and these elements, when articulated, form the back of a braincase very similar to that from Steinheim, though the estimated cranial capacity of this specimen is a little greater, around 1300 cm³. We have seen that the most modern-looking part of the Steinheim skull is its rear,

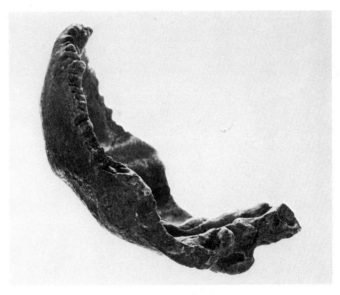

FIG. 13.2.   The Swanscombe occiput. (Two-thirds natural size.)

55

and much energy has been expended on attempts to show that the unknown parts of the Swanscombe skull would have been unlike comparable parts from Steinheim, but multivariate analysis places both squarely in the same population.

From Fontéchevade, in France, in cave deposits dating from the Riss-Würm interglacial, which lasted from about 90 000 to 70 000 years ago, have come hominid remains which in some features bear more certain resemblances to modern man. The specimens consist of both parietals and part of the frontal bone of one individual, and an isolated frontal of another. These bones are all thick, and the parietals are primitive-looking, but the frontals show no evidence of brow-ridges, and the forehead seems to have been steep. Where do these fossils fit into the evolutionary scheme? Once again, the picture appears to have been far from simple, but there is no supporting evidence for the existence of men of the modern type at this time, and the Fontéchevade fragments, with their combination of primitive and apparently advanced features, are best viewed at present as evidence of great morphological variability among late Pleistocene hominid populations.

Remaining hominids from the late Pleistocene of Europe are generally classed as Neandertals of the 'classic' or 'progressive' types. The name *Neandertal* derives from the fact that the first discovery of this type of man was made in a cave in the valley of the Neander River, West Germany, in 1856. The find consists of a calotte and a number of postcranial bones, dating from early in the last (Würm) glacial. The postcranials, though rugged, are not distinct from those of modern man, while the calotte, with a cranial capacity of about 1200 cm³, is long and low, possessing brow-ridges and a prominent occiput. This individual is of the so-called 'classic' Neandertal type; other fossils classed in this group have been found widely over western Europe, well-known finds having been made at Spy, in Belgium; La Chapelle, La

Ferrassie, Le Moustier and La Quina, in France; Monte Circeo, in Italy; and Forbes Quarry, Gibraltar. All of these sites are of early Würm date, say 70 000 to 50 000 years ago, and the fossils form a fairly homogeneous population, characterised by the features mentioned at the beginning of this chapter. Their cranial capacity is large, well up to the modern male average of around 1400 cm³; the La Chapelle skull has a capacity of 1620 cm³.

From a slightly earlier time, the Riss-Würm interglacial, come the hominids frequently styled 'progressive' Neandertals. At Krapina, in Jugoslavia, the fragmentary remains of a dozen or so individuals have been found. As far as the condition of the specimens allows us to determine, the skull is only moderately long, and the brow-ridges are not very pronounced; the back of the skull is nicely rounded. However, fragmentary though it is, the Krapina population does show a good deal of variability, and the description above is based on the skull of an adolescent. There is the further possibility, recently put forward, that the Krapina site is rather later than originally thought, perhaps dating from an interstadial during the Würm glacial. More definitely dated, but of less certain morphology, is a braincase, of Riss-Würm age, from Ehringsdorf, in Germany. As originally reconstructed, this fossil shows a relatively modern aspect; but a more recent, and more reasonable, reconstruction by Dr Pilbeam gives it a much more 'classic' Neandertal look. Finally, two Riss-Würm skulls from Saccopastore, Italy, have been described as of 'progressive' Neandertal type. It is true that the back of the skull is quite rounded, but the 'classic' features of strong brow-ridges, relatively flattened cranial vault, receding forehead, and so forth, are present.

All that the 'progressive' types do, in fact, in so far as they can be called progressive, is to emphasise the variability in skull form, especially at the back, which existed among the hominids of western Europe during the latter part of the Pleistocene. Some of this observed variation may have been

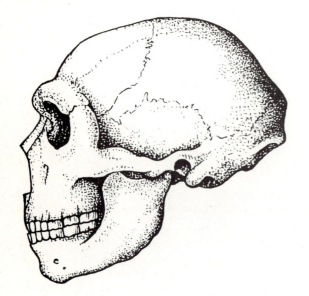 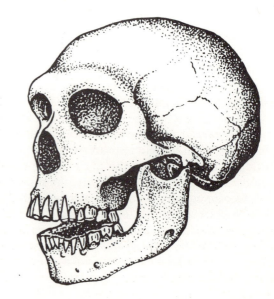

FIG. 13.3. Skull of Monte Circeo Neandertal (*left*) compared with that from La Ferrassie.

due to the migration into Europe of peoples from outside, but it is more likely that Europe during the late Pleistocene was populated by indigenous humans evolved from earlier European stocks. The 'progressive' Neandertals do not represent convincing evidence for the existence of two hominid lines at this time.

The Riss-Würm and Würm hominids of Europe are characterised by a distinctive stone-working technology known as the *Mousterian*, evidence of which is also found at the same time in Mediterranean Africa and parts of Asia. What is commonly referred to as the 'Neandertal problem' stems largely from the fact that this culture was abruptly replaced in Europe about 35–40 000 years ago by the more advanced *Upper Palaeolithic* industries, also represented elsewhere, which appear to have been the products of men of modern aspect. The inference generally drawn from this is that the Neandertals were suddenly replaced by modern types. This view, however, has long had its opponents, the most recent of whom is Dr C. L. Brace, who has very

archaic and modern types in the establishment of this population.

A similarly peripheral area is Java, where late Pleistocene deposits on the Solo River, containing a fauna known as the Ngandong, have yielded the remains of several hominids which, though the beds are not absolutely dated, are probably of Riss-Würm equivalent age. Although the term 'Neandertals' should strictly be limited to the European forms, it is useful to use 'Neandertaloid' as a general descriptive term for the extra-European peoples of approximately the same time and morphology. The Ngandong deposits have yielded 11 calvaria, with cranial capacities ranging from a little over 1000 cm³ to 1255 cm³, and two tibiae. The Solo skulls are more evolved than those of Java *Homo erectus*, but they are still decidedly primitive, having, besides relatively small brains, only moderately high skulls, with projecting occipitals, large faces, strong brow-ridges, and marked post-orbital constrictions. From a breccia cave at Wadjak, also in Java, come a couple of skulls essentially

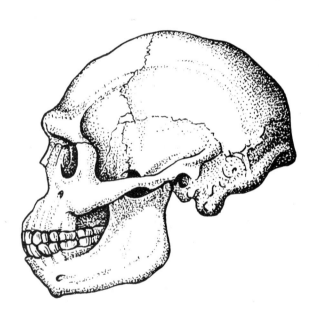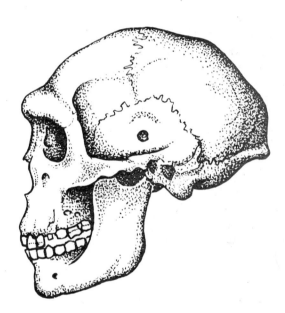

FIG. 13.4. Skulls of 'Solo Man' (*left*) and 'Rhodesian Man' from Broken Hill.

vigorously supported the theory of a gradual transition in Europe from men of Neandertal to those of modern type. There is a 20 000-year gap in Europe (from about 50 000 to 30 000 years ago) between the most recent known Neandertals and the earliest known fully modern men, and there is no evidence, as Brace is quick to point out, that the latest men of the European Mousterian were of Neandertal physical type, or that the earliest makers of Upper Palaeolithic tools were modern-looking. I think, however, that the exact sequence of evolutionary events was too complex to be unravelled from our present patchy evidence, and that though it seems likely that an outside genetic influence had much to do with the final establishment in Europe of modern man, it is probable that there was considerable gene-flow between

modern in form; they are accompanied by a fauna similar to that of Java today, and probably date from a time equivalent to the latter part of the Würm. The earliest fossil of modern aspect from this area which has been absolutely dated is the skull of a young individual from the Niah cave in Borneo, Carbon-14 dated at about 41 500 years. It has been suggested that the Solo and Wadjak skulls represent part of the evolutionary line leading to the modern Australian aboriginals. The Niah skull does not appear to form part of this sequence; rather, its affinities may lie with the aboriginal inhabitants of Tasmania, who have become extinct since the arrival of Europeans on that island.

In the upper levels of the cave at Choukoutien which yielded the *Homo erectus* material discussed in the last chapter, have been found the fossil remains of three individuals who lived in Würm-equivalent times. These skulls are considerably more advanced than their middle Pleistocene compatriots, as are other skulls from China approximately contemporaneous with them. Attempts have been made to show that hominid fossils from China all form part of the same evolutionary lineage, represented by the Mongoloids of today, but although they certainly represent sequential points in hominid evolution, it cannot be determined whether the sequence was indigenous throughout. Racial characteristics are difficult enough even for experts to discern in modern skulls; the difficulties are greatly magnified with temporal remoteness.

In sub-Saharan Africa we do have evidence of the persistence of a single lineage from the middle through the late Pleistocene, but the picture becomes more obscure thereafter. We have already seen that Olduvai Bed II 'Chellean Man' possessed brow-ridges of more impressive proportions than those of his coevals elsewhere; the same is true of most late Pleistocene hominids of sub-Saharan Africa. Particularly, the skull of 'Rhodesian Man' from Broken Hill, Zambia, is in comparable parts extremely reminiscent of Olduvai Hominid 9, although it has a larger cranial capacity, of about 1300 cm³ (modern size). A skull very similar to that from Zambia—long, large-faced, with heavy brow-ridges—is known from Hopefield, South Africa. Both these specimens may be around 40–50 000 years old. But so clear a picture of continuity is too good to be true, and its simplicity is marred by the discovery by Leakey of fragments of three skulls at Kanjera, in Kenya. The date of these skulls is not certain, but may be as much as 60 000 years ago. Reconstructed, the Kanjera crania are long, low and thick, with angulated occipitals. But the brows, in striking contrast to Rhodesian Man's, are smooth, and the faces delicate. An interpretation which regarded these differences as evidence of variability within a population would probably stretch credulity a bit too far; despite the limited extent of the evidence, it is likely that two different populations are represented here, that represented by the Kanjera material being intrusive. But where the latter came from, and what exactly are its relationships to the Rhodesian type, are unknown. After about 40 000 years ago, the Rhodesian type disappeared from Africa; no transitional types to modern man are known. What became of the Rhodesian stock? A variety of theories has been proffered: Rhodesian Man evolved into the modern Negro; evolved into the modern Bushman; became totally extinct. As we have already noted, racial skeletal characteristics are tricky enough to identify in modern man; in fossil forms, especially when, as is usual, only single specimens are available, the task becomes virtually impossible. It seems most likely, however, that the lineage of Rhodesian Man did become extinct during the late Pleistocene; but as Don Brothwell has suggested, it probably contributed genes to the incoming peoples.

In North Africa events seem to have been simpler, although this apparent simplicity may be due to the fact that hominid material from the late Pleistocene of this region is limited to mandibular and lower facial fragments. There is a trend towards the modern form, primarily consisting of size reduction, in a series of fossils covering the period from the late middle or early late Pleistocene in Morocco to about 35 000 years ago in Cyrenaica, Libya, where a jaw from the cave of Haua Fteah approaches the modern condition.

During the time of the last glaciation, interesting evolutionary events were taking place in the eastern Mediterranean area. It is here that there is evidence of the transition from the Neandertaloid type to modern man. That is not to say, of course, that this was the sole source of modern Homo sapiens, who subsequently spread from here throughout the world; evolution was obviously occurring in hominid populations all over the Old World. But it is here that the transition is best documented, and it is from this area that the people spread who probably provided the dominant genetic influence in the establishment of modern man in Europe. From the cave of Tabūn on Mount Carmel, Israel, have been recovered an almost complete skeleton and other material of late early or middle Würm age, perhaps 40 000 years old. Restored, the best skull is reminiscent of the western European Neandertal type; it is small, long and low, with brow-ridges and a strong, though flat, face. The back of the braincase, however, is rounded, and the cranial capacity is about 1300 cm³. From the cave at Shanidar, in Iraq, are known a series of deposits which cover the time period from about 100 000 years ago to the present. Unhappily, there was a hiatus in deposition between about 40 000 and 30 000 years ago. Beneath this unconformity stone tools are of Mousterian type, and are accompanied by skulls of Neandertaloid aspect. The best of these, Shanidar I, 44 000 years old, generally resembles that from Tabūn, particularly in possessing a combination of archaic features at the front of the skull with a rounded occiput. Above the break, industries are clearly of Upper Palaeolithic type; here, as in Europe, there is no direct evidence of transition. But at the cave of Skhūl, close to Tabūn, but probably a few thousand years younger in age, have been found the remains of ten individuals who almost undoubtedly represent the precursors of fully modern man in this part of the world and in Europe.

The best-preserved skull from this site, Skhūl V, has a short, high and rounded cranial vault, and a relatively flat, though robust, face. The forehead is high, but brow-ridges still remain. In sum, apart from the retention of brow-ridges, this is a completely modern skull. There has been a great deal of debate concerning whether the Skhūl and Tabūn people belonged to the same population; the most recent opinion is that of Don Brothwell, who believes that the Neandertaloid population of the area, as represented by Tabūn, was replaced by a more 'sapient' people, of whom the Skhūl sample is representative. However, there are similarities between the two populations, as well as differences, and saying that the emergence of modern Homo sapiens in any particular place is due to immigration rather than evolution is ultimately unproductive. It is interesting to note that the stone-working industries at Tabūn and

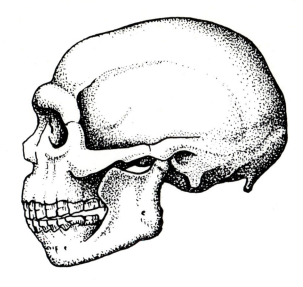 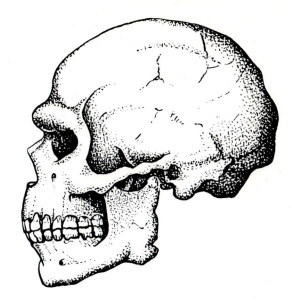

FIG. 13.5.   Tabun 1 (*left*) and Skhul V.

Skhūl are almost indistinguishable in technique, and of Levalloiso-Mousterian type.

Whether or not the Tabūn and Skhūl people represent the same population, it is fairly certain that during the time of the last glaciation, evolution to modern cranial form was taking place in the Near-Eastern area. It is also likely, as we mentioned earlier, that a similar transformation was in process in other parts of the world at about the same time, although we do not at present have the evidence to construct a world-wide picture. At the causes of this evolutionary change, and at the reasons for the success of the new type, we can only guess.

All human fossils younger than about 30 000 years are fully modern in every particular. The men of the Upper Palaeolithic are found in association with evidence of the highest achievements in the making of flint implements, and in art. But although there is abundant evidence that these Upper Palaeolithic peoples possessed highly developed and ritualised cultures (as, indeed, did the Mousterians), the economic basis of their society was essentially unchanged from that of hominids millions of years earlier. They were still hunters and gatherers. Hunting and gathering, although a reliable means of subsistence for small groups, imposes strict limits on social, economic and technological development. Again we must look to the Near East for evidence of a new departure. Around 9000 or 10 000 years ago, certain eastern Mediterranean peoples adopted a new mode of existence, characterised by archaeologists as the *Neolithic*, or New Stone Age. These people were the first agriculturalists and domesticators of animals. The change to such means of subsistence, involving the abandonment of a nomadic way of life for a settled existence, was the essential prerequisite of population expansion and the development of specialised social and economic roles, and laid the foundations upon which the complex urban and industrial civilisation of today ultimately rests.

14 Language and the Evolution of the Brain

So far we have been discussing the evolution of man's peripheral organs. In concluding, let us now consider briefly some recent research which has begun to shed light on man's *central* adaptations, those occurring in the brain itself. Language has long been recognised as one of man's most distinctive features, but unfortunately most anthropologists have tended to regard the ability to speak as an attribute of a generalised human 'cleverness'. The work of Dr Norman Geschwind, however, suggests that there is a definite anatomical basis for language; man can speak because he has evolved specialised neural equipment for doing so—equipment which only he possesses. Many authorities have tended to assume that the capacity for language depends largely on the possession of specialised peripheral vocal equipment, for instance a pharynx capable of changing its cross-sectional area, but as Dr P. C. Reynolds has pointed out: '. . . the great vocal precision manifested by contemporary speakers is the end-product of linguistic-neural evolution, not a necessary prerequisite for language.' But in the case of the central equipment the reverse is true. Human language is not merely a highly refined variant of the type of vocal communication characteristic of non-human primates; it is an entirely novel and unique system, involving the function of novel and unique neural structures.

The sequence in which different parts of the human brain mature is closely correlated with the development of the brain through phylogeny. The *primordial* zones, those which are earliest to mature, make up the major part of the cerebral cortex of a non-primate mammal. These zones include the classic *motor cortex*; the sensory regions of the *visual cortex*, the *auditory cortex* and the *somesthetic* cortex (although this last area receives sensory information from internal organs as well as from the skin, the sensations it mediates may loosely be described as 'touch'); and what may be called the *limbic system*. The limbic structures, in Geschwind's words, 'mediate both the inborn motor sequences involved in those elementary activities related to the survival of the organism or of the species, and the subjective experiences related to those activities'. The motor responses include those related to rage, fear and sexual activities, while the sensory responses include the subjective feelings of smell, taste, hunger and thirst.

Among the primates, the primordial zones comprise progressively smaller proportions of the cerebral cortex, becoming increasingly separated by phylogenetically more recent cortex which matures later and is therefore described as *intermediary*. This new cortex is known as *association cortex*, and in man comprises the major portion of the cerebral hemispheres. The area of the most striking enlargement in the human brain is the inferior parietal region, the site of the *angular gyrus* which as one of the latest parts of the brain to mature is numbered among the *terminal zones*. There is some argument over whether or not the angular gyrus is unique to man, but it is agreed that the remarkable degree of its development among humans is totally unique.

It has been recognised since the beginning of this century that in man the primordial zones effectively have no anatomical pathways between them. Each primordial zone is significantly connected only to the association cortex lying next to it. The association cortex itself, however, has long connections to other areas of the brain, though each section of this cortex may be more strongly connected to some of these cortical regions than to others. For instance, the visual association cortex has its largest connections with the lateral basal surface of the temporal lobe of the same side, which area is in turn connected to the association cortex of the limbic system. As Geschwind points out, it is reasonable that the largest connection of the visual areas should be with the limbic system, since it is obviously vital that an organism should be able to distinguish which visual stimuli are relevant to the survival of itself or of its species. Thus a monkey can be trained by positive reinforcement (reward) to choose between two differently shaped objects as long as the reward is of the type mediated by the limbic system. Since the monkey has no visual-somesthetic connections, however, having been taught to choose between shapes he can see, he cannot use this experience in choosing between the same shapes if he is permitted only to feel them. We may then say that the monkey is limited to making associations between limbic and non-limbic stimuli. Man, in contrast, is capable of forming associations between at least certain kinds of non-limbic stimuli (in particular between visual-auditory and somesthetic-auditory stimuli), and Geschwind believes that this provides the anatomical basis for the ability to understand names ('names' in this context are not limited to nouns; words such as 'fast' and 'green' would also be regarded as names), a fundamental prerequisite of language-use. In its simplest form, object-naming most frequently involves the formation of an association between sensations in the visual cortex aroused by the sight of an object, and the name of the stimulus, an association which in normal adult humans is localised in part of the auditory association area.

How does man make these associations? We have seen that the angular gyrus is effectively unique to man, and that its late maturation suggests a relatively recent origin. This

requires the liberation of vocalisation from emotion and the lowering of the stimulus level needed to elicit vocalisation. Geschwind's hypothesis provides an anatomical explanation for the fulfilment of these requirements.

Dr Jane Lancaster has suggested that a correlation may exist between the sequence of events in the ontogeny

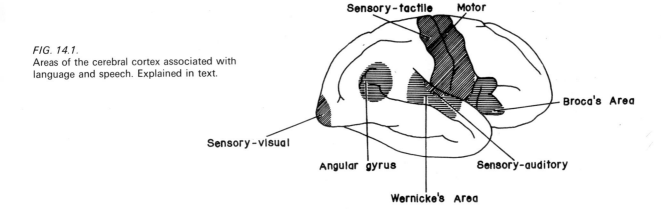

FIG. 14.1.
Areas of the cerebral cortex associated with language and speech. Explained in text.

area lies between the association areas for audition, somesthesis and vision, and is connected to all three by abundant short association fibres. It is therefore ideally situated to mediate between them, or, in other words, to act as the association area of the non-limbic sensory association areas, through which associations between non-limbic stimuli can be made.

A similar anatomical basis can be proposed for another important aspect of language: the ability to repeat words heard. The pathway in this instance goes from the auditory association cortex of *Wernicke's Area* (Fig. 14.1), through a system of fibres known as the *arcuate fasciculus*, to *Broca's Area*, the association area of the part of the classic motor cortex which initiates movement in the muscles of the peripheral vocal apparatus.

In a study of vocal and display behaviour among primates, Dr R. J. Andrew suggested that the development of language

(development in the individual) and phylogeny of language. The initial year or so of language-use among children is restricted entirely to environmental reference, mood and emotion being expressed during this period by displays— weeping, sulking—similar to those of non-human primates. A similar use of language may have characterised the earliest hominids, whom we may assume had highly developed display systems for communicating emotion. In such a context, object-naming, even of the crudest sort, would have represented a considerable advance, providing an entirely new form of communication. Only later, on this hypothesis, would language have taken over many of the functions formerly served by displays. Dr Lancaster believes that tool-use and language may have been closely interlinked, but this is uncertain; although language is surely prerequisite to many aspects of human culture, it has yet to be determined exactly which.

Selected References

General

Buettner-Janusch, J. (1966). *Origins of man.* John Wiley, New York and London.

Campbell, B. G. (1967). *Human evolution: an introduction to man's adaptations.* Heinemann, London.

Clark, W. E. Le Gros (1962). *The antecedents of man,* 2nd ed. Edinburgh University Press.

Pilbeam, D. R. (1970). *Human origins.* Thames and Hudson, London.

Chapter 1

Mayr, Ernst (1963). *Animal species and evolution.* Belknap Press (Harvard University), Cambridge, Mass.

Simpson, G. G. (1953). *The major features of evolution.* Columbia University Press, New York.

Smith, J. Maynard, (1966). *The theory of evolution,* 2nd ed. Penguin Books, Harmondsworth, Middlesex, and Baltimore.

Chapter 2

Simpson, G. G. (1961). *Principles of animal taxonomy.* Oxford University Press, London.

Simpson, G. G. (1964). 'The meaning of taxonomic statements.' *In*: S. L. Washburn (ed.): *Classification and human evolution.* Methuen, London.

Chapter 3

Eicher, D. L. (1969). *Geologic time.* Prentice-Hall, Englewood Cliffs, New Jersey.

Simons, E. L. (1967). 'Unraveling the age of earth and man.' *Natural History,* February 1967, pp. 53–9.

Chapter 4

Buettner-Janusch, J. (ed.) (1962). 'The relatives of man.' *Ann. N.Y. Acad. Sci.* 102 (2).

Jolly, Alison, (1966). *Lemur behavior.* University of Chicago Press.

Napier, J. R. and N. A. Barnicot, (eds) (1963). 'The primates.' *Symp. Zool. Soc. Lond.* no. 10.

Napier, J. R. and P. H. Napier (1967). *A handbook of living primates.* Academic Press, London.

Reynolds, Vernon (1967). *The apes.* Dutton, New York.

Chapter 5

Campbell, B. G. (1967). *Human evolution: an introduction to man's adaptations.* Heinemann, London.

Oxnard, C. E. (1969). 'Mathematics, shape and function: a study in primate anatomy.' *American Scientist,* 57 (1), pp. 75–96.

Napier, J. R. (1967). 'The antiquity of human walking.' *Scientific American,* 214 (6), pp. 56–65.

Schultz, A. H. (1969). *The life of primates.* Universe Books, New York.

Chapter 6

McKenna, M. C. (1966). 'Paleontology and the origin of primates.' *Folia Primatologica,* 4 (1), pp. 1–25.

Simons, E. L. (1963). 'A critical reappraisal of Tertiary primates.' *In*: J. Buettner-Janusch (ed.): *Evolutionary and genetic biology of the primates,* vol. 1, pp. 65–129. Academic Press, New York.

Szalay, F. S. (1968). 'The beginnings of primates.' *Evolution,* 22 (1), pp. 19–36.

Chapter 7

Simons, E. L. (1965). 'New fossil apes from Egypt and the initial differentiation of the Hominoidea.' *Nature,* 205 (4967), pp. 135–9.

Simons, E. L. (1967). 'The earliest apes.' *Scientific American,* 217 (6), pp. 28–35.

Simons, E. L. (1967). 'Review of the phyletic interrelationships of Oligocene and Miocene Old World Anthropoidea.' *In*: Problèmes actuels de paléontologie (Evolution des ver tébrés), *Coll. Int. Cent. Nat. Recherche Sci.* 163, pp. 597–602.

Chapter 8

Pilbeam, D. R. (1970). 'Miocene Pongidae of East Africa.' *Bulletin,* Peabody Museum of Natural History. Yale University, 31.

Pilbeam, D. R. (1970). '*Gigantopithecus* and the origin of Hominidae.' *Nature,* 225 (5232), pp. 516–19.

Simons, E. L. (1964). 'The early relatives of man.' *Scientific American,* 211 (1), pp. 51–62.

Simons, E. L. and P. C. Ettel (1970). '*Gigantopithecus.*' *Scientific American,* 222 (1), pp. 76–85.

Simons, E. L. and D. R. Pilbeam (1965). 'Preliminary Revision of the Dryopithecinae (Pongidae, Anthropoidea).' *Folia Primatologica,* 3 (2–3), pp. 81–152.

Chapter 9

Pilbeam, D. R. (1966). 'Notes on *Ramapithecus,* the earliest known hominid, and *Dryopithecus.*' *Am. J. Phys. Anthrop.* 25, pp. 1–6.

Pilbeam, D. R. (1969). 'The earliest hominids.' *Nature,* 219 (5161), pp. 1335–8.

Simons, E. L. (1968). 'A source for dental comparison of *Ramapithecus* with *Australopithecus* and *Homo*.' *S. Afr. J. Sci.* 64 (2), pp. 92–112.

Chapter 10

Berggren, W., J. A. Wall and L. Phillips (1967). 'Late Pliocene-Pleistocene stratigraphy in deep sea cores from the south-central North Atlantic.' *Nature*, 216, pp. 253–4.

Ericson, D. B. and Goesta Wollin (1967). 'Pleistocene climates and chronology in deep sea sediments.' *Science*, 162 (3859), pp. 1227–34.

Flint, R. F. (1957). *Glacial and Pleistocene geology*. Wiley, New York and London.

Selli, Raimondo (1967). 'The Pliocene-Pleistocene boundary in Italian marine sections and its relationship to continental stratigraphies.' *In*: M. Sears (ed.): *Progress in oceanography*, vol. 4. pp. 67–78. Pergamon, London.

Chapter 11

Clark, W. E. Le Gros (1967). *Man-apes or ape-men?* Holt, Rinehart and Winston, New York.

Day, M. H. (1965). *Guide to fossil man*. Cassell, London.

Pilbeam, D. R. and E. L. Simons (1965). 'Some problems of hominid classification.' *American Scientist*, 53 (2).

Robinson, J. T. (1956). 'The dentition of the Australopithecinae.' *Transvaal Museum Memoirs*, 9.

Tobias, P. V. (1967). 'The cranium and maxillary dentition of *Australopithecus (Zinjanthropus) boisei*.' *Olduvai Gorge*, vol. 2. Cambridge University Press.

Chapter 12

Clark, W. E. Le Gros (1964). *The fossil evidence for human evolution*, 2nd ed. University of Chicago Press.

Tobias, P. V. and G. H. R. von Koenigswald (1964). 'A comparison between the Olduvai hominines and those of Java and some implications for hominid phylogeny.' *Nature*, 204, pp. 515–18.

Chapter 13

Brace, C. L. (1964). 'Fate of the "classic" Neanderthals: a consideration of hominid catastrophism.' *Current Anthropology*, 5 (3).

Brothwell, D. R. (1963). 'Where and when did man become wise?' *Discovery*, June 1963, pp. 10–14.

Howell, F. C. (1951). 'The place of the Neanderthals in human evolution.' *Amer. J. Phys. Anthrop.* 9, p. 379.

Chapter 14

Geschwind, Norman (1964). 'The development of the brain and the evolution of language.' *Monograph Series on Language and Linguistics*, 17, pp. 155–69.

Reynolds, P. C. (1968). 'Evolution of primate vocal-auditory communication systems.' *American Anthropologist*, 70 (2), pp. 300–8.

Index